THEORY OF THOUGHT

Second Edition, 2014
Printed in Canada

ISBN-13: 978-0-9868699-4-5 (White Cover)
ISBN-13: 978-0-9868699-5-2 (Black Cover)

White Cover

ISBN 978-0-9868699-4-5

9 780986 869945

Black Cover

ISBN 978-0-9868699-5-2

9 780986 869952

This book uses QR codes. They link to Youtube videos that help clarify its text. Any smartphone or tablet with a camera can read QR codes, however you may need to download an application for your device to use this feature. If so, search your device's app store for 'QR code reader'.

For those without camera-equipped, mobile devices, the videos can be seen at:
http://www.theory-of-thought.com/videos

for one neo eon

TABLE OF CONTENTS

PATTERNS

CHAPTER I

CONCEPTS

SYMBOLS

PHILOSOPHY CHAPTER IV

ATTRACTION CHAPTER V

PHYSICS CHAPTER VI

MECHANICS CHAPTER VII

PREFACE

The idea for this book struck me on November 21st,
2006. I began writing hours later on November 22nd.
My mind fell into a euphoric state with feelings
of an epiphany. The sensation lasted for about
a week, until November 29th, and to this day, I have
considered it my moment of enlightenment. It was
a strange state of mind that produced an unstoppable
rush of ideas. It felt like my mind was rewiring itself,
because a single idea touched upon everything I knew.
Soon enough, my thoughts were re-arranged into
a new paradigm. It has taken me over six years
to work through my paradigm shift and to describe
it concisely. This book is my way of remembering it,
studying it and sharing my vision of it.

January 6th, 2013

I've written Theory of Thought to share a new worldview and to find like-minded people that can assist in expanding it and unleashing its full potential.

*The difference between making
a breakthrough and not
can often be just a small
element of perception.*

— Brian Greene

INTRODUCTION

As people think, events unfold - decisions are made, doors are opened, and things are moved. It seems that thought motivates people to act as if they were pushed by some force. Over time, thought gathers into complex streams of thoughts, and from their combined forces, systems are forged, organizations are founded, and societies are built. In essence, our thoughts manifest the material arrangement of our physical world.

As the popular saying goes, our lives are 'mind over matter', but how can that phrase be explained scientifically?

Our traditional framework for explaining how thought affects matter has remained the same for millennia. Generally, when discussing the way in which people interact with objects, we use the following simple model: the brain commands the body and the body manipulates matter. Very little attention is paid to the energy within thought, and as a result the traditional framework doesn't explain how symbolism might ultimately drive the displacement of matter. The following theory will expand on the notion of thought and matter by exploring the concept of a hyperdimensional space that unites the brain, body, and mind into a single logical entity.

Theory of Thought synthesizes the physical world with the abstract world of symbolism using a new paradigm. It assembles a unique model that explains energy, light, gravity, motivation, love, and other driving forces as consequences of hyperdimensional organizations interacting through an invisible region of space and time that is embedded within everything around us.

I think the universe is more profound than we commonly believe and that it extends beyond a collection of bits of matter in a 4-dimensional spacetime. I believe that reality may be better defined according to a geography of minds rather than a geography of bodies. I will also argue that the reality of nature has more to do with symbolic forms of structure, as described by **metaphysics**, rather than physically tangible ones. Metaphysicists believe in abstract forms that exist beyond the material world, as elaborated by Plato's Theory of Forms. If this notion were proven true, the composition of the Universe may be shown to be less materialistic, and more abstracted than we commonly believe. In this book, I shall illustrate that the Universe is designed to contain symbolic forms of structure within a hyperdimensional space, while matter and thought are the means by which we experience them.

This book is about thought; but not in any conventional way. It will explain how a thought is an abstraction for a real structure which exists within a higher dimensional region of spacetime. The Universe is hyperdimensional because these symbolic structures have more than 4 dimensions. This book argues that a brain is but a complex organization that evolved to coordinate the motion of symbols within an abstract and invisible region of the Universe. So where exactly in the Universe do these symbolic structures reside?

I've always been intrigued by the notions within string theory, and M-theory. At its root, string theory predicts that the dimensions of space go beyond the traditional **physical dimensions**. The traditional physical dimensions are **height** (up and down), **width** (left and right), **depth** (front and back), and **time** (travel across space). This book will refer to these **spacetime** dimensions, as the 'physical dimensions', or the 'physical space'. String theory expands upon the physical dimensions because it predicts the existence of some extra dimensions, which it calls 'higher dimensions'. String theory proposes that the Universe contains at least 6 more infinitesimally small dimensions that are folded into each other at every point in space. It should be noted that these dimensions have never been observed and are therefore purely abstract. By including these abstract dimensions within equations essential to modern physics, the force of gravity can be reconciled with quantum mechanics - perhaps solving a long standing issue in physics. As a consequence of its potential for scientific advancement, the study of these extra dimensions has become more broadly accepted within the physics community.

If string theory is right, it means there are at least 10 or more dimensions of matter that are omnipresent. The higher-dimensions are basically interwoven with the height, width, depth, and time dimensions - meaning that as one walks down the road, his body and every atom within it travels through both the 'physical' dimensions and some 'higher' dimensions.[1] One can visualize each bit of matter being folded (or unfolded) through these extra dimensions as any object moves through them. The net result is that there are two perspectives. The first one is from the physical space where the body is perceived as having a distinct, physical shape, as a three-dimensional entity moving across time. The second one is a comprehensive hyperdimensional perspective, where the body is a fluctuating field, bending through the entirety of a hyperdimensional space, with a structure far beyond the observed three-dimensional entity.

Theory of Thought posits that an intangible higher dimensional space, called **mindspace**, exists in the Universe. In this theory, called **thought theory**, 'hyperdimensional space' or 'mindspace' refers to an all-encompassing space that includes and extends beyond spacetime. This book's particular model of hyperdimensional space provides an interconnected theory similar to that of string theory, while explaining itself quite differently. According to the various string theories, the exact number of higher-dimensions that may exist is still unclear. For all we know, these dimensions only exist within mathematical equations and it might be that the higher dimensions may never be observed, entirely because they are 'abstract', in contrast to being 'physical'. Therefore, I have approached the notion of higher-dimensions from a different perspective, and illustrate them in logical and philosophical contexts. My arguments take these notions beyond any contemporary theory by demonstrating how the fundamental properties of thought emerge from these higher-dimensions.

Thoughts are bound to us in a way that may be more fundamental than our bond to matter. I'm writing a series of books to prove that the Universe was designed to contain abstract structures, called **symbols**, and to show that matter is but a 'physical' perspective of these invisible structures, while thought is their 'abstract' perspective. I also believe that physics is in need of a philosophy that explains the rational for the construction of the Universe, because the divide between empirical evidence and thoughtful presupposition has grown too far to ignore any longer. The gap must be bridged with a scientific theory that unites science and popular belief using a metaphysical description of mind and matter.

Where do minds exist in the Universe? What does the structure of a symbol look like? And how do these structures interact in relation to our laws of physical science? Are symbols affected by gravity for instance? I will address these questions by presenting a series of visual diagrams that I believe explain how the Universe fundamentally works.

By the end, this book will reveal a **Fundamental Mechanism** founded upon symbolism, semiotics, and mathematics. It is a universal cycle that reveals how symbolic organizations interact within a mindspace shared by all living beings. The cycle forces order upon the symbols across mindspace, by forcing upon their building blocks, called **patterns**. Symbols are collections of patterns that navigate across physical and abstract dimensions of mindspace simultaneously and ultimately, construct the world we experience. Note that within this book, I have carefully reframed many commonly known concepts to help establish my final conclusions. Most reframed words are in bold, and their respective definitions can be found in the glossary in the back of this book. In doing so, I hope to establish a scientific framework supporting how all physical arrangements of matter can be better understood as basic patterns within mindspace.

I'd like to give you an example on how this theory applies. Consider that our everyday world is organized according to products, brands, and prices, to name but a few divisions. These intelligent constructs are assembled over time as real structures, but atop invisible planes of reality. There is an entire realm of abstract structure that is organized when people think similarly and work together for common meanings. This abstract world of symbols becomes super-imposed onto the material objects around each one of us. Each object and thing exists in physical space with a height, width, length, and time, but it also exists in an abstract space that holds some form of its meanings and symbolic relationships across the man-made hierarchies of mindspace. For example, each establishment such as a home, is built and supported by the minds that think about it. There is a fundamental law of the universe that says if people are thinking about something, and if enough thought concentrates together within the abstract space, some material arrangement will take shape in the physical world to reflect that intangible structure of thought and symbolism. Thought and matter are bound within a universal cycle, with arrangements of thought acting upon arrangements of matter through the actions of symbols. And if the symbolic structures shrink in size and disappear within mindspace, so will their reflected arrangements of matter in our physical space. Therefore, the arrangement of symbols we grow and maintain in our minds manifest the world around us.

So in contrast to the conventional brain-over-body paradigm, Theory of Thought shows that complex arrangements of symbols take shape within a hyperdimensional space, that they exert themselves onto each brain (and produce thought), and that they contain mathematical properties that alter the distribution of material objects within our physical world.

I dedicate this book to my mom, who has patiently supported my journey. Thank you to all my colleagues and friends who have helped me complete it. Special thanks to Marc Kandalaft for designing my work along the way. I also thank the Wikipedia organization for publicly sharing such a terrific source of information. I would not have reached my goals without all their wonderful contributions.

God does not play dice with the Universe

— Albert Einstein

PATTERNS
CHAPTER I

RELATIONSHIP

CIRCLE
container

LINE
pathway

FIVE PATTERNS

Building blocks of mindspace

**HIERARCHICAL
PATTERN**

**UNION
PATTERN**

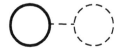

**NETWORK
PATTERN**

**SYSTEM
PATTERN**

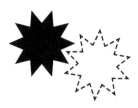

**GEAR
PATTERN**

The following theory is entirely founded on these 5 'relationship' patterns. Each concept it presents is described using one or more of these basic building blocks of mindspace. Years ago I asked myself what built our world, and my answer was *relationships*. Everything can be connected and related, and I've found that these 5 patterns provide the exact manners in which relationships form an invisible web that interconnects all our bodies, thoughts and minds.

Ref. page(s): 198 **THEORY OF THOUGHT**

NUMEROUS CONCEPTS

Illustrations of anything

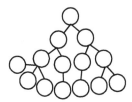

HIERARCHIES
Found in trees, family trees, charts, pyramids, operating systems, cell-division, and the general concept of acceleration.

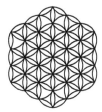

UNIONS/VENN
Found in plants, sacred geometry, irrational numbers, contracts, religion, community, and the general concept of symmetry.

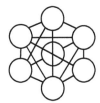

NETWORKS
Found in distribution grids, the Internet, maps, and the general concept of motion.

SYSTEMS
Found in colonies, societies, organs, galaxies, fractals, security systems, and the general concept of individuality.

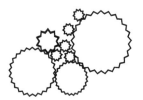

GEARS
Found in machines, engines, clocks, robotics, astrolabe, electron interactions, and the general concepts of work and friction.

The 5 patterns provide the contextual features of all living and non-living arrangements of matter. It is presumed that everything in the universe, including thought, can be fully illustrated and measured using these relationship diagrams.

Thought theory hypothesizes that the structure of thought and symbols must be visualized in a logical and geometric nature in order to discover the true architecture of a **mind**. This chapter will show that symbols have distinct features, which a mind employs to interact with a material environment. My view is that a mind constructs its abstract reality of symbols by employing exactly five relationship patterns. These **5 relationship patterns** configure the space of a mind into **hierarchies**, **systems**, **networks**, **unions**, and **gears**. As a result, any person, object, or organization in the world can be fully described using these five types of patterns. Furthermore, there is a hidden mind-matter balancing mechanism that sustains arrangements of matter, in parallel to their arrangements of patterns in mindspace.

In short, all material forms in our environments are visibly and invisibly arranged into exactly five types of relationships. People had to think to produce all these arrangements of objects: buildings, cars, roads, signs, furniture, electronics, etc. These material arrangements are imbued with thought, emitted by symbolic patterns within higher dimensions of reality. All forms of matter have stored values of 'meaning', and thus each is inherently more complex than physical reality portrays. This theory is an examination on where hidden value resides. How do symbols gather power and energy? I believe the answer is found in examining the intersection of physics and psychology: Imagine if every thought you had, extended into an invisible realm that somehow impacted the visible world? Well in thought theory, matter and meaning is interwoven into a unified structure called a symbol, which is bound to a shared mindspace of relationships and governed by probabilistic laws of nature.

Note that mindspace has 3 fundamental divisions, comprising of exactly 12 dimensions:
1) 4 spacetime dimensions that *construct* patterns into material shapes.
2) 4 intellectual dimensions that *define* complex arrangements of patterns.
3) 4 wave dimensions that *equilibrate* patterns and material shapes.

In this section I will go over the basic patterns that permeate a mindspace. Each pattern type is founded on the symbolic meanings of the circle and line.

CIRCLE
A basic container

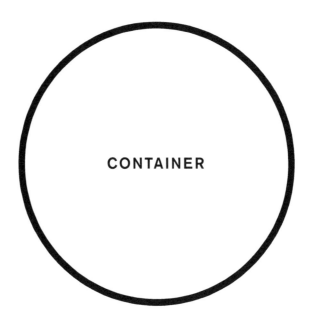

CONTAINER

Philosophically, the **circle** describes a single, united container. It represents a void, a unity, an organization, a cycle, and a completion.[1] In metaphysics, it is the ideal state of creation and the most important of all known shapes. For instance, our Sun is quite simply a circle of fire in the sky, that since the dawn of time has directly fueled the creation of thought in all living creatures on Earth.

THOUGHT

Emission of substance

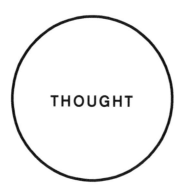

THOUGHT

ANY
ORGANIZATION

In mindspace, a circle is a meaningful organization shaped from **thought**. Any worldly object can be encompassed by the ideal shape of the circle, transformed into a symbol, and manipulated by a mind.

ORGANIZATIONS

Symbols of function

THOUGHT OF ANY BUSINESS

outer-organization

MANAGEMENT
DIVISION

MARKETING
DIVISION

inner-
organizations

SUPPORT
DIVISION

SALES
DIVISION

ACCOUNTING
DIVISION

MANUFACTURING
DIVISION

SYSTEM

Circles contain smaller circles called **inner-organizations** that are divided in thought and symbolism. Inner-organizations are separate containers of energy that support or give life to their shared container called the **outer-organization**. The whole organization may be that of a company, or perhaps that of a human body. For instance, within a human body the first layer of inner-organizations are commonly called 'organs'.

SYSTEM

Boundary

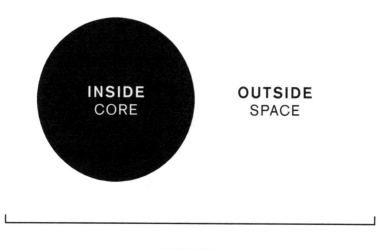

INSIDE
CORE

OUTSIDE
SPACE

SYSTEM

SYSTEM

The circle is the prototypical enclosed system. As a shape, the circle requires an *inside*, an *outside*, and a *border* separating its inside and outside. Scientists call this concept a 'system' and it can help model other concepts ranging from biological beings, to pockets of hot air, to the nucleus of an atom.

PATHWAY

Line between boundaries

PATHWAY

NETWORK

In contrast to the circle, the **line** is disconnected; it has disjointed end points. Instead of being a circle, the line is given a distinct purpose: to bridge circles. In metaphysics, the line represents a bridge, a pathway and a distance. It can be a road between cities or some relation between any two points. Circles, thoughts, organizations, systems, symbols and arrangements of matter are various words for explaining the same, basic phenomenon: the organization of complex structures of circles and lines within an abstracted hyperdimensional space.

COOPERATION

Building mindspace

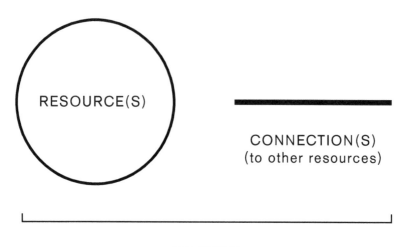

RESOURCE(S)

————

CONNECTION(S)
(to other resources)

PATTERN

NETWORK

In thought theory, the complete notion of a symbolic organization requires two structures: the circle and the line. The circle stores symbolic resources that have value, while the line transfers them to other organizations. In a physical sense, the line can be a pathway that exchanges something tangible. While in an abstract sense it can be a pathway that exchanges something intangible. For example, when two people are speaking to one another, they are connected by a line that exchanges physical and abstract forms of resources such as heat and information, respectively.

CONNECTIVITY

More than one connection

NETWORK

Symbols can have large numbers of connections that interact with many other symbols simultaneously. Some of their connections can travel long distances into far away regions of the mindspace. These lines could represent the exchange of any moving, energetic object. Notice how the illustration looks a lot like the symbol of the Sun, surely because the Sun maintains the most connections of any single object in our solar system.

RELATIONSHIP

Two systems sharing values

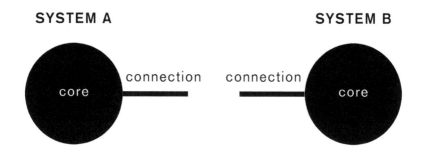

SYSTEM A   **SYSTEM B**

core   connection   connection   core

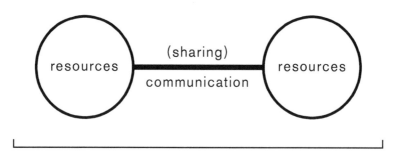

resources   (sharing)   resources
communication

RELATIONSHIP

NETWORK

All systems are forced into relationships with other systems. Even systems that seem disconnected must in fact be connected across some distance. In thought theory, distance is abstracted into a concept called **symmetry**. According to the rules of symmetry, symbols that are similar in meaning will gravitate in mindspace and expand their connectivity. Through symmetry, symbols communicate and exchange energy according to thought theory's Fundamental Mechanism; I will explain this mechanism in greater detail later in the book.

SYMMETRY

Values being exchanged

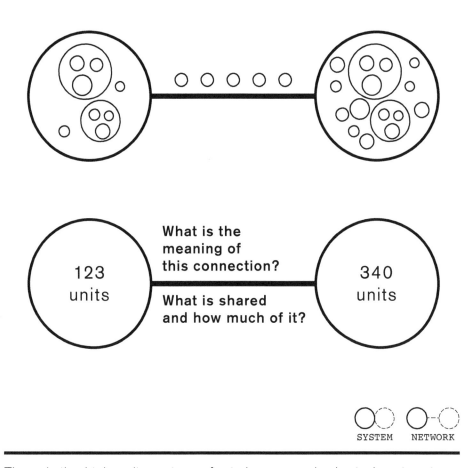

123 units	**What is the meaning of this connection?**
	What is shared and how much of it?

340 units

SYSTEM   NETWORK

Through the higher dimensions of mindspace, each physical system tran-scends into a hyper-dimensional symbol that holds some abstract value. Its value is a function of its complex arrangement, which in turn exerts an attractive force upon external symbols in order to build increasingly complex relationships. The exact values of the exchanged resources can be compli-cated to measure as the values fluctuate when patterns are re-arranged.

NETWORKS

Relationships fuel symbols

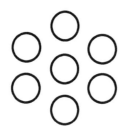

DISCONNECTED SYSTEMS

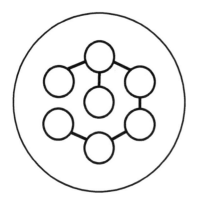

INTERCONNECTED SYSTEMS

SYSTEM   NETWORK

Inner-organizations are bound into a whole organization that is greater than the sum of its parts. Disconnected organizations, that aren't sharing resources, cannot survive within an outer-organization and they will always eventually disappear from it. The relationships fuel progress while growing the lifespan of the outer-organization. So any organization, or society, that appears to be to be healthy and growing contains a well connected group of supporting organizations.

COMPLEXITY

Measure of connectivity

ADD

SIZE TO

THE ORGANIZATION

=

ADD CIRCLES

low interconnectivity
low complexity

ADD

COMPLEXITY

TO THE ORGANIZATION

=

ADD LINES

high interconnectivity
high complexity

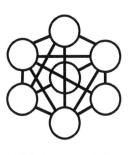

NETWORK

Complexity drives the survival of each organization in the group. Every organization maintains a balance between its size and complexity known as its **complexity ratio** (henceforth **c-ratio**). **Size** is the number of its inner-organizations and **complexity** is the number of its inter-connections. The c-ratio is like the fingerprint of the organization, and is instrumental in explaining a symbol's communication pathways across its arrangements of patterns. The equation for the c-ratio is: $\dfrac{\text{\# of lines}}{\text{\# of circle}}$

ANCIENTS

Tree of Life, ancient

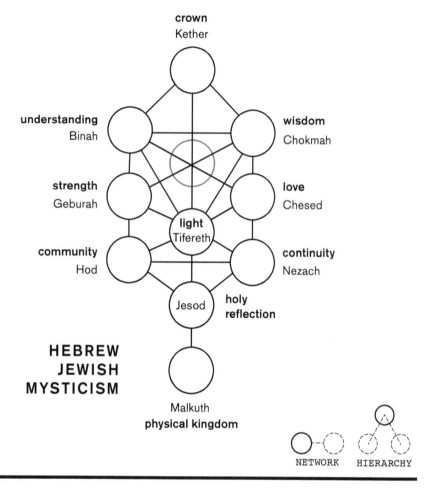

crown
Kether

understanding
Binah

wisdom
Chokmah

strength
Geburah

love
Chesed

light
Tifereth

community
Hod

continuity
Nezach

Jesod

holy
reflection

HEBREW
JEWISH
MYSTICISM

Malkuth
physical kingdom

NETWORK   HIERARCHY

Ancient Pythagorean geometers and philosophers believed in a logical **hierarchy** defined by the **four dimensions of intelligence: meanings, shapes, numbers**, and **letters**. All minds naturally access these dimensions to identify and categorize symbols. These abstract dimensions lead to intelligent patterns that can be deciphered only by other minds. In the figure above, the outer-organization is called the Tree of Life, and the relationships between its ideas of love, strength, and wisdom have been purposefully shaped into the design of a tree.

THEORY OF THOUGHT

TREES

Hierarchical organizations

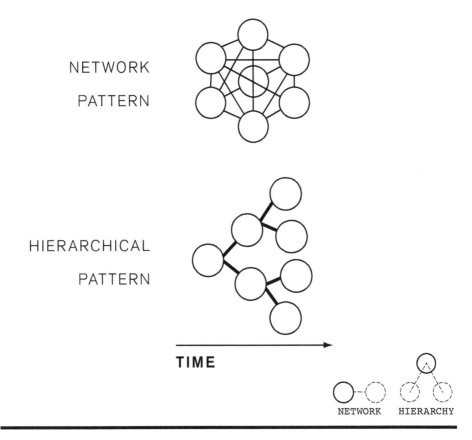

NETWORK

PATTERN

HIERARCHICAL

PATTERN

TIME

NETWORK   HIERARCHY

Tree diagrams help illustrate ancestry, similarity, and division. They are hierarchical patterns that explain complexity with respect to time. Not only do organizations communicate and share resources, but there is a hierarchy of communication and sharing that grows and stretches across time. In mindspace, new symbols are assembled into the branches of these trees, while older symbols provide the fuel for growing new branches of hierarchy. The hierarchical pattern is significant because it is perhaps the most important blueprint used for the arrangement of matter in nature. Using this basic blueprint, all people rearrange matter into increasingly ordered hierarchies, which leads to progress, invention, and civilization.

PERSPECTIVES
Viewpoints of complexity

TOP VIEW | SIDE VIEW

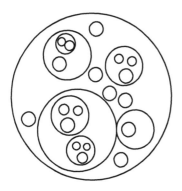

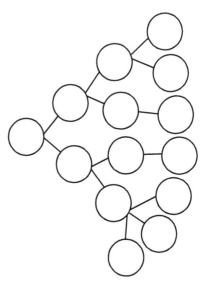

SOCIETIES | TREES

SYSTEM

HIERARCHY

Organizations can be viewed from different perspectives. The *system pattern* helps explain organizational sub-division. A *network pattern* helps describe organizational communication, and the *hierarchical pattern* illustrates its ancestry (layers in mindspace). Understanding these patterns will help us understand thought, and how thoughts are emitted from a hyperdimensional space of patterns.

DIAGRAMS

Complexity of organizations

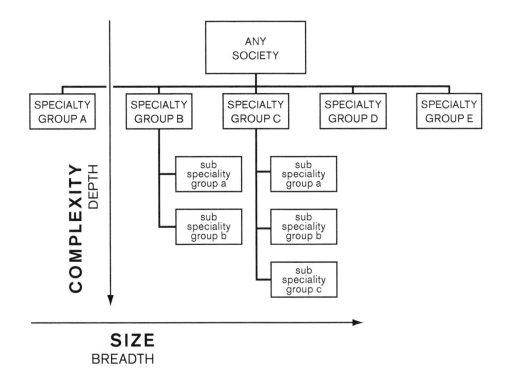

HIERARCHICAL DIAGRAM

HIERARCHY

Every complex organization can be mapped using a similar diagram. The *complexity* component reveals new *branches*, while the *size* component fills branches with new *resources*. Organizations use their energy-distributions to interact with neighboring organizations. Organizations with the most effective organizational patterns in any given environment will appear to out-compete others, because they are better equipped at establishing and maintaining relationships.

DARWIN'S TREE

Tree of Life, 1863 AD

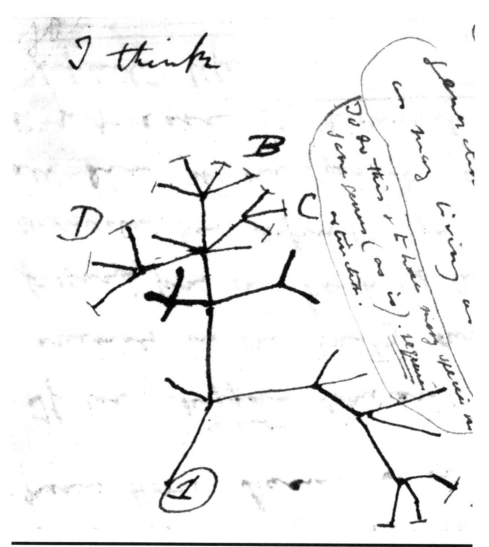

Darwin was the first scientist to explain that organizations evolved over time, using natural selection. He described his theory using a tree-network diagram that visualized populations of species expanding and dividing from one another as time forces the differentiation of new species from their predecessors. Can Darwin's theory of evolution also apply to structures of thoughts, symbols, and all worldly organizations in general?

NEURONS

Hierarchical patterns

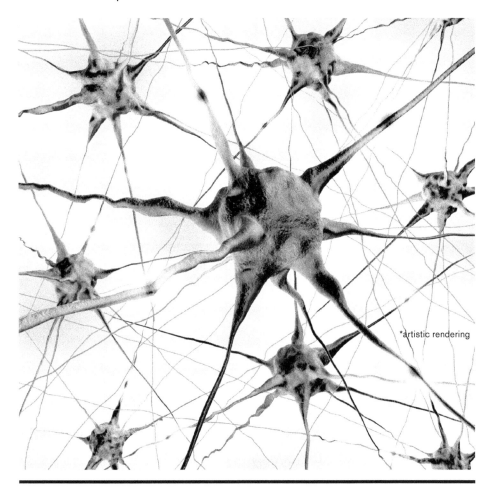

*artistic rendering

The neural arrangements in the brain are known to order information using the principles of systems, networks, unions, mechanical, and hierarchical patterns. Basically, the brain is a network of interconnected patterns that are sensed from the physical world. Its purpose is to evaluate the hidden complexity of material arrangement and translate it into 5 types of patterns which the mind stores in mindspace. When a brain is unable to convert what it senses into well-organized patterns, a person will prioritize the wrong relationships with his or her environment and may inadvertently lower his or her value from what appears to be poor decision making.

UNITS OF THOUGHT

Simplistic nature

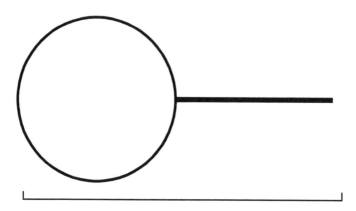

PATTERN - Basic building block

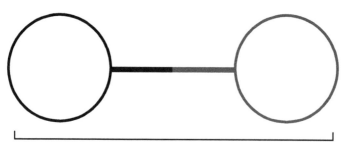

SYMBOL - Contains more than one pattern

NETWORK

The brain assembles the sensory information it receives into symbolic units and it encodes them across a relational mental map that contains 12 distinct properties (dimensions). Resources flow between related symbols, and if some resources fail to reach important units, stacks of symbols will collapse, neurons will stop communicating, and thoughts will evaporate.

ASSEMBLY

Creating ordered sequences

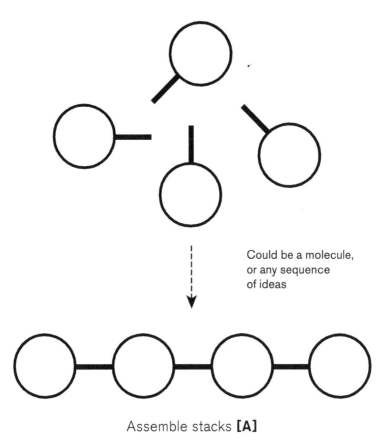

Could be a molecule, or any sequence of ideas

Assemble stacks **[A]**

NETWORK

Through an abstract force of nature, called **attraction**, separated, yet related units join into arrangements that I call **stacks**. An abstract force of nature controls symbols, as well as matter. In fact, systems and organizations of all scales are under the influence of attraction, forcing them to assemble into **complex arrangements** within mindspace. It should be said that life and evolution are forced by attraction 'behind the scenes'.

Patterns within patterns

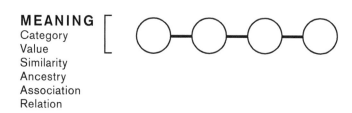

MEANING
Category
Value
Similarity
Ancestry
Association
Relation

- -

Stacks of thought combine into strands of thought, building complexity.

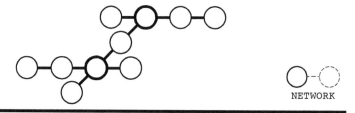

NETWORK

The principle feature of the living brain is that it translates arrangements of matter into relationship patterns. The brain is a translation device between the physical world and the metaphysical world.[2] With respect to Kant's theories, the unconscious brain is capable of mapping all the hidden relationships present within an environment. Thought theory states that the unconscious mind builds this map by using the five relationship patterns of mindspace. Within the unconscious, the patterns are under the control of some fundamental laws of mindspace, namely attraction. Consciousness arises as the brain's focal point on a region of the mindspace map. During consciousness, the symbolic patterns of mindspace come to life and project thought into the brain. Each symbol has 12 distinct dimensions: meaning, shape, number, letter, period, amplitude, frequency, wavelength, height, width, depth, and time. And minds can be found at every **scale** of the Universe thinking with respect to these specific 12 dimensions. The questions arise: How do minds connect patterns together? Why do they connect? When, where, which, or whose minds are connected? The answers to these questions will help explain how minds interact across mindspace, forging physical reality.

MINDSPACE
Complex sets of relationships

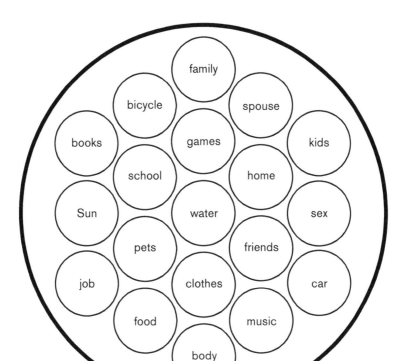

THE MIND

SYSTEM

All minds inhabit an *abstract* region of spacetime, while the brain inhabits its *physical* region. My mind is the complete 'thought of me' and it only contains relationships. For example, it comprises of the relationships I have with the objects I own, the people I know, the things I've sensed, and so forth. In its hyperdimensional environment, a mind can be influenced, supported, or hurt by other minds, and its actions drive physical reality by forcing living bodies

MIND AND BRAIN

Shared symbolism

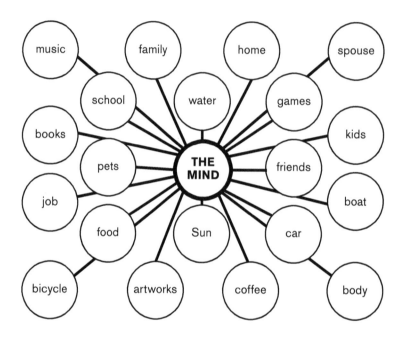

NETWORK

Since the mind and the brain are different entities, I will clarify each of their roles. The brain is a material organ that dies with a body, and it functions as a complex management center for minds. As a brain is assembled, a mind flows into it. During a lifetime, the brain coordinates the construction of the mind and a mind will continue affecting the real world long after a brain that developed it has disappeared, because a mind is an assembly of patterns that have been successfully ordered together into symbolic organizations across an abstract space.

From the perspective of your brain and its consciousness, you have ideas about your family, however, you are not directly connected to your family. You are a separate entity with its own disconnected existence. That being said, from the perspective of your mind and the unconscious, your being is innately tied to that of your family, because of the relationships you share with them. Individuality is a grey area within the mind's world, since the mind is a network of relationships. And because a symbol can maintain many relationships simultaneously, minds can appear to have significant overlap and war for control over symbols in mindspace. Furthermore, as minds navigate and interact with these patterns, people and objects move across physical space replicating their behaviors.

Ultimately, each brain and body is a physical reflection of a metaphysical mind that inhabits and navigates mindspace. As such, the mind is naturally in tune with the goal of mindspace: building, maintaining, and breaking patterns of relationships.

Nature's grand book, which stands continually open to our gaze, is written in the language of mathematics.

Its characters are triangles, circles, and other geometrical figures, without which it is humanly impossible to understand a single word of it; without these, one is wandering around in a dark labyrinth.

— Galileo Galilei, 1623

CONCEPTS
CHAPTER II

ENERGY

TIME
potential

SPACE
kinetic

It's one thing to realize that the Universe is driven by a framework of hidden patterns to which each mind has access, but it's another thing to present a coherent mathematical argument that explains the congruence of patterns with all widely held scientific laws of nature. This chapter sets the stage for such a proof with a series of conceptual pillars, from which thought theory will eventually extend.

The Universe is a network of energy structures in various forms, such as matter, motion, and light. Energy can take several physical forms, however, is the symbolism found within a logo or photo a form of energy? I will argue that symbols contain abstract energy that is sustained in mindspace using the concepts of the circle and line. Furthermore, these two primordial shapes will be shown to be the fundamental building blocks for all values of energy (physical and abstract).

Here are some facts about **energy**:

> 1 — Energy exists between the dimensions of Time and Space.
>
> 2 — Energy can be in a potential state (rest), and/or kinetic state (moving).
>
> 3 — Mass is energy. Energy is mass.
>
> 4 — Across time, all masses of energy fluctuate.
>
> 5 — Interactions cause fluctuations in energy, which increases entropy.

In thought theory, energy is stored within each mindspace pattern. Minds manipulate these patterns, and thereby move bodies of interconnected matter across spacetime using their energies. Minds are themselves 'living' patterns and they are instinctively attracted to outside patterns that contain large amounts of energy, since they are sustained by these potential resources. One basic way to understand mindspace is to imagine a web of travelling circles and lines. Our goal is to understand what factors drive the construction of the web of circles and lines, and the rules that guide the motion of circles across lines.

I believe that the rules guiding the arrangement process are deeply related to our innate notions of love. The pre-socratic philosopher Empedocles once defined love as a force that attracts the four classical elements[1], because love has always been considered to be a pulling force akin to gravity, or a derivative of the warmth of light. In thought theory, love, motivation and energy can be unified into a single framework based entirely on the simplicity of these two geometric shapes.

The circle and the line are partners in a duality that permeates everything. The Universe and the structures that are assembled within it are rooted in their basic symbology. Together, they create the fundamental mechanism of mindspace and they are its two *fundamental perspectives*. To understand this better, consider the following: from the top, a circle looks like a circle, however, from the side a circle looks like a line. Through a single shift in 90° perspective, between top and side views, the basic geometry of the 2-dimensional circle can be seen as the foundation of mindspace.

Take note that the circle and line are *unique* types of regions in mindspace that are differentiated only by perspective.

Consequently, these basic shapes recur in all forms of nature, and through the flow of time and the creativity of thought these two shapes are repeatedly arranged together, particularly with respect to motion and communication:

Information is often stored as circles and lines, known as binary code.

Prior to modern information systems, Morse Code employed dots and dashes and was one of the principle methods of communication.

Mechanical motion is best achieved by using wheels (O) and axles (I).

'Zero' Point Energy is a quantum theory. So is 'Loop' Quantum Gravity. So is 'String' Theory. All of these names contain references to circles or lines.

The Ancient Egyptian stories of Isis and Osiris are among some of the oldest sources of religious symbolism.

The list of coincidences between these two shapes is extensive and this simplistic partnership often goes unnoticed. So how do we continue *connecting the dots* and how *deep does the rabbit hole* go?

Over the next two chapters I will overview a series of concepts that will be key for understanding the final chapters. Each idea will be presented to you like a conceptual tool, each with a unique function. Once I've introduced all the tools, I will re-construct the abstract machinery that builds symbols.

TIME AND SPACE

Web of patterns

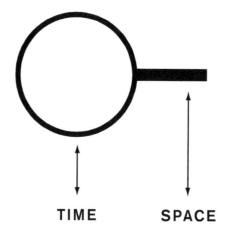

TIME   **SPACE**

A whole unit of the network pattern represents the scientific framework of 'spacetime' (ie. time and motion). In thought theory, the circle represents time, and the line represents space. From this assumption, time can be basically thought of as a rotation like that of a clock, while space is thought of as linear motion like that of a travelling object. Throughout this book I will present many arguments for these exact associations. One particularly compelling argument is that each one of us inhabits a circle of time, interacting with one another across the lines of space. This idea follows Einstein's theory on the independency of time for separately moving observers. This diagram re-interprets relativity theory and lays the foundation for visualizing how the measurement of a symbol's time can be affected by the lines it maintains.

LIVING FORCE

Mechanical energy

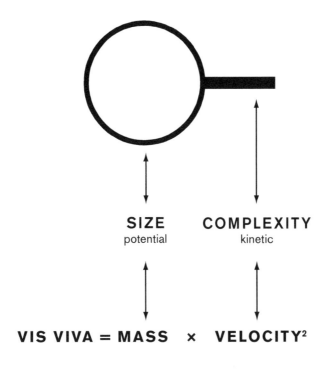

SIZE
potential

COMPLEXITY
kinetic

VIS VIVA = MASS × VELOCITY²

NETWORK

The circle and line should be understood as the two basic perspectives for every block of energy in the Universe because each shape represents a unique state of motion: potential and kinetic. **Vis Viva** is the Latin phrase for the 'Living Force', and it was coined to describe mechanical motion. In thought theory, it is hypothesized that the equation describing the Living Force also applies to the energy in symbols and thought. Minds manipulate the potential and kinetic energy found within abstract patterns, by manipulating the 'complexity' (c-ratio) of such patterns.

MOTION

Rotations and straight lines

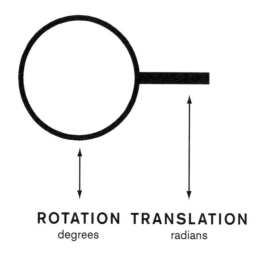

ROTATION **TRANSLATION**
degrees   radians

NETWORK

There are exactly two ways in which something can move:

O – An object can be rotated or twisted to form a circular movement.

I – An object can be pushed or pulled to form a linear movement.

Rotations (O) and translations (I) are the only two basic types of motion. This entire pattern, containing both rotation and translation, represents a unit of energy. Energy is the stuff that travels the pattern in the figure above, or more accurately, this pattern *is* energy. Energy is not a mysterious phenomenon and it can be logically visualized as a basic building block of symbols in mindspace.

ARC AND LINE
Measuring energy

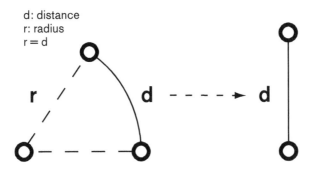

DEGREES
Observer's viewpoint

RADIANS
Mover's viewpoint

Arcs and straight lines measure distance from
two unique perspectives: the mover and its observer.

NETWORK

Energy is measured through distance. In thought theory, distance is reframed as symmetry, and its used to calculate the separation between two symbols in mindspace. Like distance, symmetry can be measured in conventional units (meters), but it can also be measured in any standard. In mindspace, it relates abstract distances - such as one between a boy and a man. This distance might be the difference in the number of jobs each has held. It might also be used to measure the 'distance' between a couple of teenagers in love. This might be the sum of love letters shared between them over a period of time. I believe there should be standardized rules for measuring all forms of symmetry based on the same rules for measuring distances. However, how do we go about measuring all the degrees and lengths of personal relationships?

TIME

A cycle of energy

clock

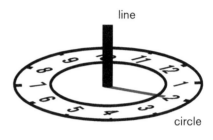

line

circle

sun dial

All clocks can be used to measure the displacement of energy over time. For instance, one might register a force of 1 kg m per second squared when measuring mass, but if measuring the strength of *a marketing campaign*, one might register 10 people per minute. In particle physics, energy travels in small organizations called quanta, and each one is valued using both space (*1 kg*) and time (*per second*). In thought theory, the force of 10 people per minute is similarly stored within quantized organizations of time, called patterns.

Furthermore, it has been shown that the motion of matter results in an effect called time dilation. Time dilation forces one system to age faster than another, and it causes time travel. In his special theory of relativity, Einstein explained that time is a relative experience, meaning that the flow of time across my organization and the flow of time across your organization could be different if we moved at very different speeds. In thought theory, it is hypothesized that a form of abstract time dilation could be taking place between symbols. A change in c-ratio might result in the **time dilation of mindspace**, causing a noticeable environmental shift that might manifest as the displacement of an additional X people per minute through a door. If abstract time dilation can be linked to the behavior of living bodies, it would reveal that the primary task of the mind is to manipulate the cycles of time within a hidden region of space.

Ref. page(s): 33   **THEORY OF THOUGHT**

CONSTANT π

Concentration

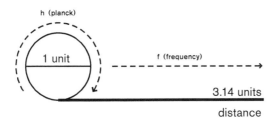

h (planck)

1 unit

f (frequency)

3.14 units

distance

π

3.14159…

**IRRATIONAL
TRANSCENDENTAL
INTELLIGENCE / ATTRACTION
CONCENTRATION**

One rotated line
One concentration

pi: 'perimeter' (greek)
peri-: 'to measure'

pi is the ratio between the
concentric circumference and the
linear diameter of a circle.

SYSTEM

Generally speaking, **π** is the most important constant in the Universe. **Pi** is an irrational number because it cannot be represented as a fraction - the decimals of pi extend randomly and non-periodically into infinity. However, there are stone tablets from Ancient Babylon and Egypt, circa 2000 BC, that have the oldest known references to its value: 22/7. [2] Interestingly, in Ancient China and India, π was believed to be the √10 (ie. a square root of a circle and line). Its value is used in many of the most important equations that describe the universe, such as Euler's Identity and Einstein's formula for the energy of a photon. In thought theory, the constant of π is reflective of a force of rotation that permeates mindspace to create units of time.

ROTATION

Generating waves

CONCEPTS OF ROTATION

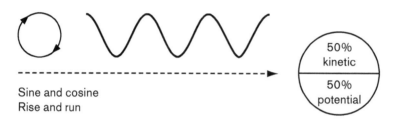

Sine and cosine
Rise and run

50%
kinetic

50%
potential

CONCEPTS OF TRANSLATION

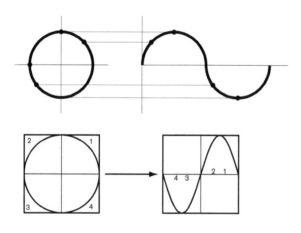

SYSTEM

A **wave** is a periodic disturbance transferring energy across space and time and it is said to contain 50% **potential energy** and 50% **kinetic energy**.[3] Waveforms can be reproduced by uniformly rotating a circle.[4] Sine and cosine are pillars of geometry because they relate lines, circles, waves and rotations into mathematical equations. In mindspace, there are rotating circles that travel the web of lines, and as they rotate, they reflect waves back into our physical world.

WAVE

Dimensions of a waveform

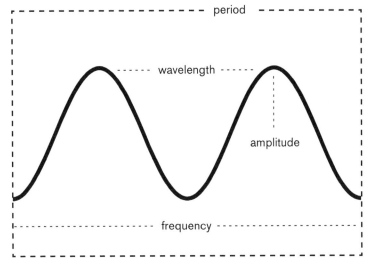

PERIODIC WAVE

reduced
planck's
constant

$$\hbar = \frac{h}{2\pi}$$

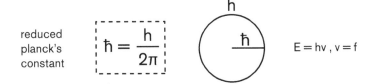

$E = h v \, , \, v = f$

SYSTEM

A periodic wave contains exactly four properties. These four properties are of paramount importance in thought theory because they are considered to be the **wave dimensions** of mindspace: **period**, **amplitude**, **frequency**, and **wavelength**. All cycles and waves in nature are manifested by the uniform and non-uniform periodicity of circles within hyperdimensional space. As a consequence, the basic equations at the foundation of quantum mechanics can be easily related back to circles.

OSCILLATION

Rotation with translation

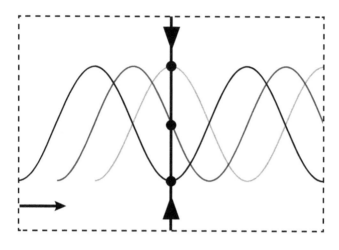

SIDE VIEW

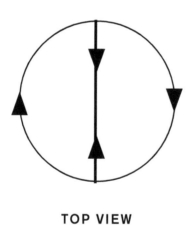

TOP VIEW

Richard Feynman suggested that the universe could be fully described in terms of harmonic oscillators.[5] A simple harmonic oscillator is a type of wave, moving in periodic, sinusoidal repetition with a constant amplitude. In physics, the activity of the electromagnetic field is measured by calculating a series of oscillators, and it is hypothesized that each symbol in mindspace can also be described as an oscillator.

CONSTANT *e*

Scaling and growth

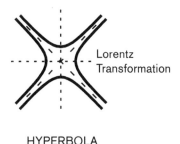

Lorentz Transformation

HYPERBOLA

$e \approx 193 / 71$

2.7182818284590...

1 828 1 828

palindromic

**IRRATIONAL
TRANSCENDENTAL
WAVE / NEGATION
NATURAL LOG BASE**

2.71828...

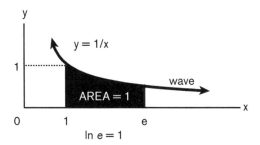

logarithms
scaling
multiplication
growth
instantaneity

The constant *e* is a prominent sign of calculus, the branch of mathematics that measures waves. This constant is often referred to in its natural logarithmic form - *ln*. In thought theory, *e* is tasked with guiding the exchange of energy between symbols in a natural, harmonic method, and as a consequence, its value helps grow and shrink their rotating diameters.

SPIRAL MOTION

Motion of sub-atomic particles

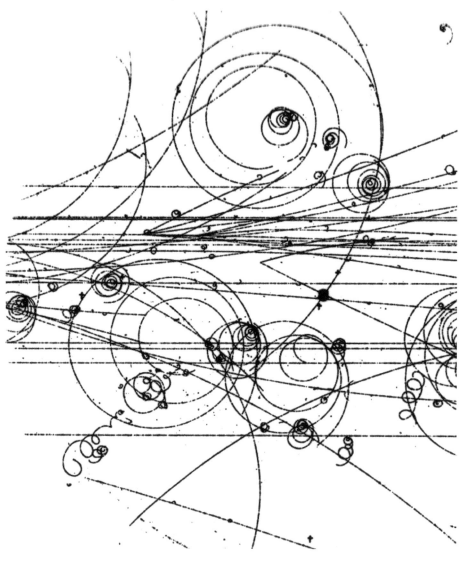

The constant *e* is often associated with the logarithmic spiral. A spiral is a composite shape that arises from circular rotation undergoing linear translation. This is a photo of the motion of sub-atomic particles across the fabric of spacetime. [photo[6]]

PERIODIC TABLES

Small scale rotation of space

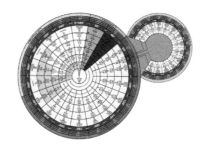

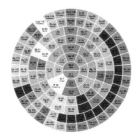

ALTERNATIVE PERIODIC TABLES OF ELEMENTS[8]

By definition, a period is a cycle, which is a rotation. In 1862, the first periodic table of elements was developed using cylindrical diagrams. When they were first charted, scientists believed the atomic elements were formed according to some spiraling phenomenon.[7] Scientists have since decided that the step-like periodic diagram that we are most familiar with today was best suited at categorizing the elements. However, many alternative diagrams have arisen over the years to illustrate other features of the periodic (ie. cyclical) elements.

SPIRAL GALAXIES

Large scale rotation of the Milky Way

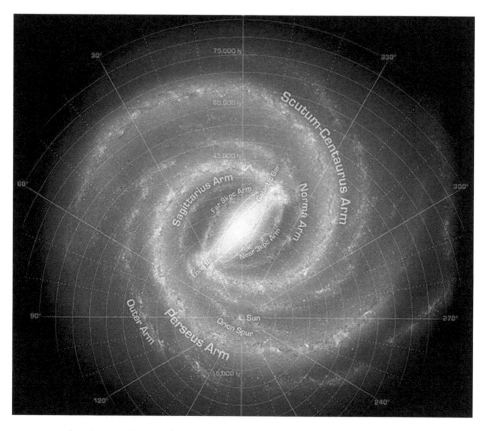

Our Sun is located within ORION'S spur [image⁹]
ORION: "Heaven's light"¹⁰

Spirals are naturally associated with growth, formation, gravity, and time. The brain is estimated to contain 10^{11} neurons and our galaxy contains about 10^{11} solar systems, therefore can the functions of brains and galaxies be somewhat compared? Do solar systems interact by sending light across deep space, like neurons interact by sending electrical impulses across synaptic gaps? Given its sheer size, it seems quite impossible to travel outside of our own galaxy. Is this a metaphor for one's inability to travel outside of their own mind? Does the spiral idealize the shape of a mind?

THEORY OF THOUGHT

A cloud is made of billows upon billows upon billows that look like clouds. As you come closer to a cloud you don't get something smooth, but irregularities at a smaller scale.

— Benoit Mandelbrot

MANDELBROT FRACTAL

An icon of the fundamental mechanism

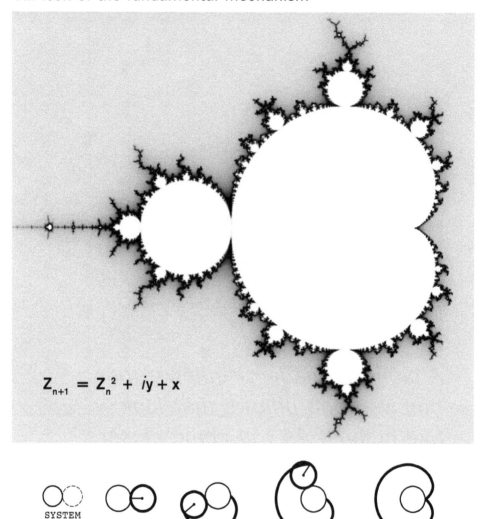

$$Z_{n+1} = Z_n^2 + iy + x$$

SYSTEM

The Mandelbrot set is a fractal first seen in 1980 at IBM. It's the simplest quadratic polynomial that can unfold across 'a complex plane'. Imagine that a complex plane is like a rotating surface - anything that is placed on it can reach the other side, not by travelling straight across it, but by rotating around with it. Using the complex plane, the Mandelbrot set is formed by having one circle rotate around another. Now imagine that every symbol in mindspace billows out into one of these fractals.

TRIANGULATION

Rules of right triangles

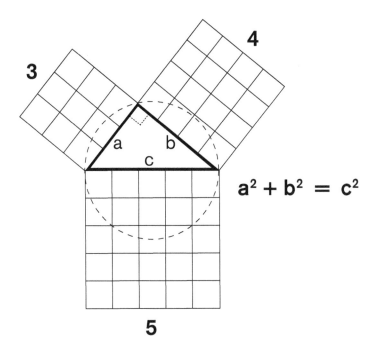

$$a^2 + b^2 = c^2$$

SPECIAL EQUATIONS

$$1^2 = (\cos x)^2 + (\sin x)^2$$

$$c^{i\Phi} = a + bi$$

$$1^2 = \sqrt{\varphi}^2 + \varphi^2$$

The **Pythagorean theorem** is one of the oldest and most important equations discovered. This equation is so important that it is found in all forms of mathematics that govern geometry, algebra, calculus, and physics. For instance, it is instrumental in calculating time dilation. It also has a special relationship with rotation and the circle. I will argue that the Pythagorean theorem, and its alternate equations, can be used to measure all forms of symmetry because our minds naturally employ it to assemble patterns within mindspace's geometric framework.

CONSTANT Φ

Cornerstone of physics: symmetry

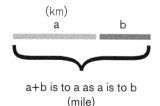

(km)
a   b

a+b is to a as a is to b
(mile)

1 mile ÷ 1 km ≈ Φ

-a justification for truncation will be
provided near the end of the book

PHI

**THE GOLDEN RATIO
IRRATIONAL
AUTOMORPHIC
CONSTRUCTION**

(Phi) φ: 1.618033...

(phi) φ: 0.618033... / 61.8%

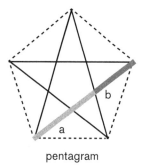

pentagram

ΦΙ...   **LOVE**
ΦΥ...   **NATURE**
ΦΩ...   **SYMMETRY**

-ancient greek associations with letter Φ

Pythagoras was fascinated by all irrational constants and **Φ** (pronounced -*fi*) may have been his most celebrated since he found it to have a mysterious connection with the pentagram and the number 5. Phi is a remarkable constant because its coincidences with other numbers, rational and irrational, are entirely unique in nature. In mindspace, **Φ** is used along with π and *e* to connect patterns into complex, yet stable arrangements. Later in the book, I will present some never-before-seen intersections between these 3 special constants.

ARRANGEMENT

Phi represents physical nature

For centuries, Φ has been referred to as the golden ratio. People have observed the golden ratio in DNA, trees, plants, and the human body, among many other arrangements found in nature. This ratio produces a noticeably attractive arrangement of simple complexity. I argue that Φ is forced into mind-space patterns during the fundamental mechanism, because it is the corner-stone of construction. While π is a ratio used to concentrate patterns into containers (circle), Φ is a ratio used to connect containers together (line).

Φ SPIRAL

Series of squares

PHI SPIRAL

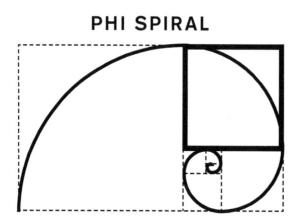

SQUARES OF PHI

$\varphi = 0.61803..$

$\Phi = 1.61803..$

$1^2 = \sqrt{\varphi^2 + \varphi^2}$

pythagorean theorem

$\Phi^3 = \sqrt{5} + 2$

$\varphi^3 = \sqrt{5} - 2$

$\Phi = (\sqrt{5} + 1) \div 2$

$\Phi^2 = \Phi + 1$

$\Phi^2 = \Phi + \ln e$

$\varphi \approx (\pi \div 4)^2$

Phi is the junction between several elementary concepts, and therefore, it expresses many functions simultaneously. Phi is a ratio; it's a wave; it's a spiral; and it relates deeply to squares. Its value, 1.618.., is the point of perfect mathematical symmetry. It is the natural binding mechanism that produces pathways between symbols, therefore its ratios, spirals, and waves are various perspectives on the functions of pathways (lines) in mindspace.

Stabilizing systems

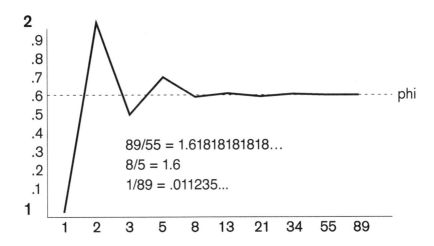

89/55 = 1.61818181818…
8/5 = 1.6
1/89 = .011235…

FIBONACCI'S SERIES
A+B=C; B+C=D,…

0, 1, 1, 2, 3, 5, 8, 13, 21, 34, 55, 89, 144…

First 12 numbers of the infinite set

HIERARCHY

Fibonacci's infinite series is calculated by A+B=C, B+C=D, … Dividing B/A, C/B, D/C, and so forth reveals ratios. Plotting these values reveals a wave that is similar to a dampened harmonic oscillator, and over time the pendulum stops overs the number 1.618… Shown above, Φ emerges from an incredibly simple series of arithmetic used by symbols to stabilize their pathways.

PASCAL'S TRIANGLE

Binomial expansion

PASCAL'S SERIES
A+A=B; B+B=C,...

PASCAL'S SERIES

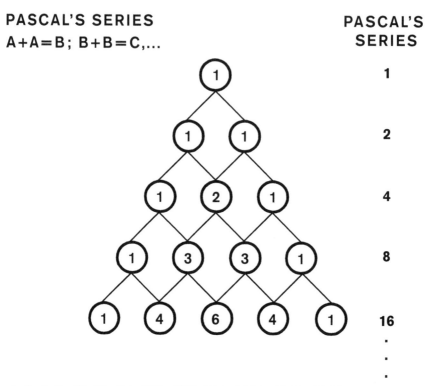

1

2

4

8

16

.
.
.

1, 2, 4, 8, 16, 32, 64, 128, 256, 514, 1024, 2048, 4096

First 12 numbers of the infinite set

HIERARCHY

Pascal's triangle was known to the ancients, because it's a peculiar root into the inner-workings of the mind. The numbers in the tree are probabilities, and I believe they are used to assemble hierarchies in mindspace. Mathematically, Pascal's triangle is a triangular array of the binomial coefficients and it is closely related to Fibonacci's series. Coloring only the odd numbers in Pascal's triangle reveals a fractal called the **Sierpinski triangle**. Thought theory hypothesizes that the process of symbolic evolution exploits the concepts within this 2-D pyramid.

FIBONACCI'S TREE

Automorphic evolution

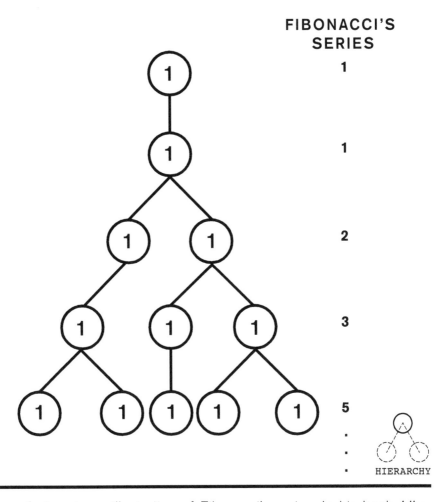

FIBONACCI'S SERIES

1

1

2

3

5

.
.
.

HIERARCHY

Fibonacci's tree is an illustration of Fibonacci's series. In his book *Liber Abaci* written in 1202 AD, Leonardo of Pisa, called Fibonacci, described an arithmetic sequence in which each number is equal to the sum of the previous two. He demonstrated that this incredibly simple series of equations is deeply related with Φ by plotting a type of evolutionary tree. He visualized the growth of an idealized rabbit population and imagined that a newly-born pair of rabbits, one male, one female, are put in a field. Rabbits are able to mate at the age of one month so that at the end of its second month a female can produce another pair of rabbits. Populations of rabbits never die and a mating pair always births one new pair (one male, one female) every month from the second month on. Fibonacci asked, how many pairs will there be in one year?[12] Using a family tree, he calculated that the rate of growth for each generation of rabbits was accelerating by Φ. He became convinced that reproduction across the universe, was guided by sacred mathematics. Since his discovery, people have argued that his example with rabbits is flawed since it depends on incest and immortality. However, I wonder if his story can be better applied to symbols in mindspace, instead of rabbits. Do symbols reproduce and spawn new symbols, and can the growth of their populations in mindspace be correlated to Φ?

I believe that the resurgence of sacred symbolism during the Enlightenment can be attributed to Phi. During this period, Φ experienced a growth in popularity because of its incredibly fascinating features that are nothing short of spectacular. Entire rooms can be filled with books containing diagrams, equations, and examples illustrating the intellectual properties of Φ (φ).

Through all the associations outlined in this chapter, I have shown that Φ is conceptually rooted in hierarchy, symmetry, structure, reproduction, evolution, growth, branching, and acceleration. These concepts can be traced back to π, *e,* lines, waves, circles, spirals, squares, and the Pythagorean theorem. In turn, those concepts can be traced even further back to the general ideas of motion and time within spacetime. All of these basic concepts are the related, foundational notions governing patterns in mindspace and when explained in a particular manner, can be tied together into a single process called the fundamental mechanism - a mathematical cycle that creates abstract objects of value, called symbols, that push and pull on the minds and bodies of everyone on Earth.

We have learnt that the exploration of the external world by the methods of physical science leads not to a concrete reality but to a shadow world of symbols

— Arthur Stanley Eddington

SYMBOLS
CHAPTER III

SYMBOLIC
BUILDING
BLOCK

SIZE
stability

COMPLEXITY
instability

A religion is a system of symbols which acts to establish powerful, pervasive, and long-lasting moods in men

— Clifford Geertz

History remembers Pythagoras as a pioneer of mathematics, cosmology, and numerology. He lived between 570-490 BC and during his lifetime he was considered the wisest philosopher in Greece. He was responsible for teaching many early Greeks about numbers and mathematics. He devoted his life to uncovering the connections between geometry, religion, and philosophy. As a reflection of his commitment to knowledge, Pythagoras popularized the word *philosophia* to describe his 'love of wisdom'.[1]

It is believed that he was taught geometry by the Egyptians, arithmetic by the Phoenicians, astronomy by the Chaldeans, and the formulae of religion and practical maxims for the conduct of life by the Magians. People in antiquity thought that Pythagoras had travelled extensively, and had visited Egypt, Arabia, Phoenicia, Judaea, Babylon, and even India, collecting all attainable knowledge, and especially that concerning the secret or mystic cults of the gods.[2]

To this day, Pythagoras is commonly associated with the triangle and a body of thought called 'Sacred Geometry'. At his core, Pythagoras was a meta-physicist who believed strongly in a very symbolic diagram called the *Tetractys*. He even founded a school devoted to worshipping it using a belief system carefully arranged between science and religion. His early philosophical designs inspired many of the future's great thinkers, including Socrates, Plato, and Aristotle. Carl Huffman said this about Pythagoras:

"Pythagoras believed that numbers had personalities. They were masculine or feminine, perfect or incomplete, beautiful or ugly. He considered ten to be the ideal number, [summed by] one, two, three, and four, and when written in dot formation it formed a perfect triangle, named Tetractys. He believed that four signified nature, and that ten signified the universe, [and the sum of 1,2,3, and 4 was divine]. The **Tetractys** is both a mathematical idea and a metaphysical symbol that embraces itself, the principles of the natural world, the harmony of the cosmos, the ascent to the divine, and the mysteries of the divine realm. This symbol was so revered that it inspired Pythagoras' followers to swear by his name as the one who brought this gift to humanity."[3]

Pythagoras was a highly respected intellectual with many remarkable ideas. It could be possible that he possessed an acute awareness for the symbolic interrelation of numbers, shapes, and their meanings. Under his views, symbolism is conceptually imparted by the Universe. In other words, symbolism cannot be invented - it has always existed. This lends support to the idea that an alien race from a distant world would have a remarkably similar symbology to ours since they must use the same numbers and shapes as we do. Circles, lines, triangles, rectangles and hexagons should be interpreted by all intelligent beings all over the Universe, in the same ways. It should be that the Star of David, the shape made from two opposing triangles, is a powerful symbol across many if not all intelligent civilizations. Like the Ancient Egyptians worshipped our Sun, all alien races surely worshipped their Sun sometime in their own history, and that in itself may have exposed their minds to the same psychological seed of symbolism that was introduced to our own minds by our own Sun.

According to Pythagoras, a shape inherits value from the meanings of numbers, and through this relation, shapes become symbolic things that deeply affect minds. Everyday, people interact with combinations of simple shapes. For example, the Internet is a network of binary shapes: 1010110. Too complex to easily recall, we hide these shapes behind the scenes, yet the meanings of 0 and 1 are everywhere, and as opposing symbols they have an effect on all of our behaviors. You probably already know this, but symbols are often used to control minds - which ultimately control bodies of people.

People are attracted to symbols, because symbols give them energy, and motivation. Forms of matter that contain letters, numbers, shapes, and meanings have a strong effect on human behavior. As symbols themselves, people need symbolic energy to grow and survive. Symbols have a real mechanism that influences the behavior of living beings by providing them with a sense of value, purpose, and direction. This chapter will explain symbols as objects of intellectual design that are rooted in very basic notions which stretch across thousands, millions, and billions of years - since the dawn of the Universe itself - ie. the birth of the first circle.

ICONIC SYMBOLS

Astrology and religion

Monad
God

Sun Pyramid
Polytheism

Pisces
Christianity

Earth
Female

Venus
Female

Moon
Female

Mercury
Male

Mars
Male

Sun
Male

This theory defines a symbol as a form of both matter and thought. Symbols are structures defined by relationships and symmetries provide them with strength and a lifespan. Powerful symbols will replicate across time much like humans, and successfully reproduced symbols are usually among the simplest in design and will have easily recognizable associations.

SYMBOLS OF JESUS

Very deeply rooted symbology

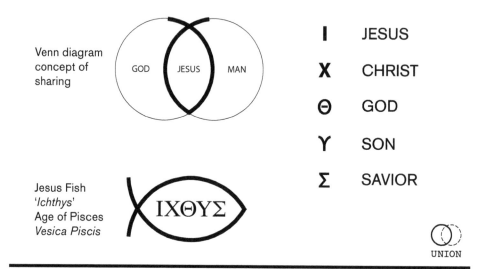

Venn diagram concept of sharing

GOD | JESUS | MAN

Jesus Fish
'*Ichthys*'
Age of Pisces
Vesica Piscis

ΙΧΘΥΣ

Ι   JESUS

X   CHRIST

Θ   GOD

Υ   SON

Σ   SAVIOR

UNION

The story of Jesus is perhaps the most communicated piece of information in history. The reason that it has been so well received and shared between people is because of its powerful, symbolic associations. Building valuable symbols requires the alignment of simple ideas, thus the story of Jesus is often related with fish and coincidentally was written during the Zodiacal age of Pisces. In these stories, Jesus speaks deeply about the value of sharing, an act relating to the Union pattern. He's often compared with the celestial realm, such as being the 'Son' (Sun), being born on December 25th (end of Solstice declination), having been met by 3 kings (Orion's Belt), and choosing 12 disciples (astronomical zodiac).

These associations supply life to the 'mind of Jesus' - a mind that lives in mindspace, interacts with other minds and symbols, and reproduces itself across time. The churches, crosses, and bibles that appear around us are there because of our proximity to powerful symbols contained within the mind of Jesus. When its symbolic associations gain in strength (popularity), a new church may be built, or perhaps its followers will appear to wear more crosses. Symbols compete for minds in mindspace and as a symbol's influence fades, its worldly reflections, like churches and crosses, will slowly fade away with it.

ZODIAC
Transcending symbols

The zodiac is a celestial and ecliptic coordinate system and its origin is from the Greek *zōidiakos kuklos*, meaning 'circle of animals'. It appears that this exact arrangement of twelve primary symbols has survived the test of time, vis-a-vis other competing astrological symbols, because it contains the strongest symbolic associations. The competing systems may have become obsolete thousands of years ago from having a lack of symmetry between its designs and the actual symbols that underpin mindspace. Symmetry should be considered as a degree of likeness between two or more things or thoughts.

INTELLIGENCE

Four abstract dimensions of symbols

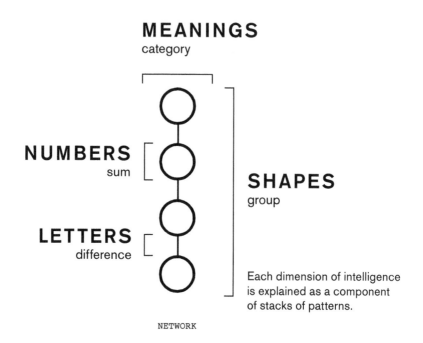

MEANINGS
category

NUMBERS
sum

SHAPES
group

LETTERS
difference

Each dimension of intelligence is explained as a component of stacks of patterns.

NETWORK

NETWORK

Look around and you will notice something: almost everything is branded with letters, numbers, shapes, and meanings. These are natural components of symbols and are interwoven into the fabric of the hyperdimensional space.[4]

I propose that people did not invent the concepts of letters and numbers. These concepts explain patterns intellectually and emerge from the architecture of mindspace. We increase our control over minds in a mindspace by developing our application of letters, numbers, shapes, meanings and combinations thereof. A strong education system improves our ability to connect symbols across mindspace thereby adding to our own symbolic value and strength.

GLYPHS

Abstract associations in material arrangement

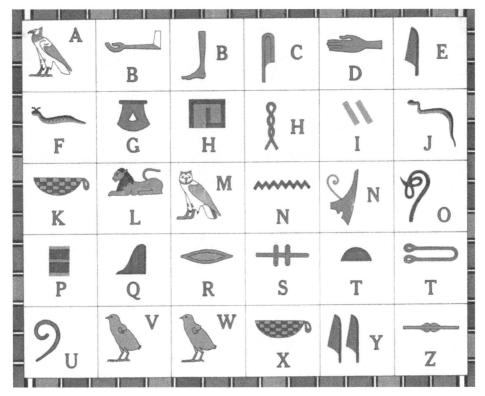

ANCIENT EGYPTIAN HIEROGLYPHS[5]
Mix of meanings, shapes and letters

A mind categorizes a symbol according to meaning, and each symbol contains many meanings. The meanings influence the configuration of patterns in a mind, and since we all share the same symbols, our minds become patterned similarly. Like all minds, alphabet and numeral systems are forced to evolve over time, and the most used letters and numbers are those which contain the strongest associations (symmetry).

NUMBERS

Numerology

N°	MEANINGS
0	Omni. Everything. Absoluteness. All. Birth. Monad. Unity. Stability.
1	Individual. Action. Yang. Monad. Male. Instability. Pathway.
2	Balance. Union. Receptive. Yin. Dyad. Female.
3	Motion. Interaction. Neutrality. Triad. Harmony.
4	Stability. Creation. Tetrad.
5	Action. Restlessness. Split. Pentagram. Half of decad (10).
6	Growth. Reaction/flux. Responsibility. Half of dodecad (12).
7	Thought. Consciousness. Number of Numbers.
8	Power. Acceleration. Sacrifice.
9	Highest level of change. Finality.
10	Rebirth. Decad. The Universe. Perfection.
11	Pathway between rebirth and balance.
12	A balance in pathways.

Numerology is an instinctual concept that explains associations between numbers, letters, shapes and meanings. Each mind assembles a numerological system because each mind is forced to organize symbols according to the four dimensions of intelligence. Commonly used numerological systems can be useful for gaining perspective over the fundamental concepts of the Universe. I find it very intriguing to find logical overlaps of the numerological beliefs of the Pythagorean, Brahmin, and Ancient Chinese societies. It appears that their meanings for the numbers between 0 to 4 are very similar. For example, these cultures agreed that the number 1 is related to an action, while the number 2 is related to restoration and balance. It's also interesting that ancient cultures regarded an odd number as male, and an even number as female. I believe that the meanings of one symbol can help us learn more about its other inter-connected symbols.

LETTERS

Containers of meaning

A = **A**ttraction
E = **E**nergy
I = **I**ntelligence
O = **O**rganization
U = **U**niverse

'the **U**niverse is an **O**rganization of **I**ntelligent **E**nergy led by **A**ttraction'

I believe the meanings of each letter are hiding in plain sight. I believe that the letter *A* looks like a compass tool used to draw a circle, and this is very significant of what *A* means. The association between the letter *A* and the triangle should also be significant. I also believe that there may be a way to decipher the meanings of letters to reveal the existence of a universal cypher that minds use to establish complex symbols within mindspace.

INFORMATION

Intersection of dimensions

**INTELLIGENCE
DIMENSIONS**

**PHYSICAL
DIMENSIONS**

Hello! Can we meet tomorrow at 6pm?

**WAVE
DIMENSIONS**

Written documents contain large assortments of repeating symbols, such as letters, signs, words, phrases and punctuation. Through symbolism, an entire document exists within 12 dimensions and its overarching meaning is correlated with the meanings of the symbols and groups of symbols it contains. For example, a book like the Bible will inherently contain a greater frequency of words with religious significance than say Euclid's Elements, which is a book on geometry.

I believe that a higher understanding of symbols has taken this long to come to fruition because symbolism is incredibly broad. It is very difficult to know where to start when examining such abstraction. It is my view that a theory on symbols should arise from the circle and line, which represent the base of symbolism. From there on, one can begin assembling a theory describing the nature of complex arrangements of meaningful shapes.

I will argue that symbols are interwoven into the structure of space and time, meaning that they can be scientifically described using models and experiments. I think that they are geometrically tied into the spacetime architecture using the Pythagorean theorem and the right triangle.

There is at least one organization of people that have already made the conscious connection between the construction of symbolism and the shape of the triangle, and by aligning itself with this meaning and shape, it has been able to reap significant benefits.

Freemasonry is a fraternal organization that arose from obscure origins in the late 16th to early 17th century. Candidates for regular Freemasonry are required to declare a belief in a Supreme Being. In the ritual, the Supreme Being is referred to as the Great Architect of the Universe, which alludes to the use of architectural symbolism within Freemasonry.[6]

Freemasonry has often been called a secret society. However, it is more correct to call it an esoteric society because only certain aspects of it are commonly understood. Two of the principal symbolic tools always found in a Freemason lodge are the square and compass. Some lodges and rituals explain these tools as lessons in conduct: for example, that Masons should "square their actions by the square of virtue" and to learn to "circumscribe their desires and keep their passions within due bounds toward all mankind." Freemasonry uses the metaphors of operative stonemasons' tools and implements, against the allegorical backdrop of the building of King Solomon's Temple, to convey what has been described by both Masons and critics as "a system of morality veiled in allegory and illustrated by symbols."[7]

Given their profound symbolic associations and their rise to prominence over business and politics since the Enlightenment era, is it possible that Freemasons are concealing knowledge about their ability to manipulate mindspace through meaning?

FREEMASONRY

Symbol of construction

FREEMASONRY
Great Architect Symbol

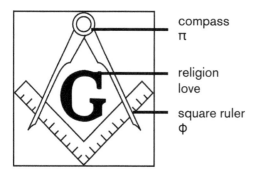

compass
π

religion
love

square ruler
φ

GREAT SEAL
Symbol of USA

HIERARCHY

In the Freemason logo, *G* represents the Great Architect, or God. The compass represents the *circle* and the square ruler represents the *line*. Furthermore, the compass intersects the ruler at exactly Φ, or 1.61. By employing these powerful, symbolic values, Freemasons have gone on to successfully out-compete other organizations. Consider that the United States of America, with a GDP that is almost 3 times that of any other nation on Earth, is founded on Masonic symbolism. Success appears to come to those that embrace mathematics, or in the very least, embrace logic.

Ref. page(s): 70   **THEORY OF THOUGHT**

PYRAMIDS

Symbols of divinity

A pyramid is a basic pattern of construction. The Λ shape is a symbolic form designed to breach into the depths of the mindspace, where Gods, as Universals and symbolic minds, do exist. Notice how the Freemasons use both the Λ shape and the pyramid to symbolize design, construction, and spirituality.

CONSTRUCTION
Lines of symmetry

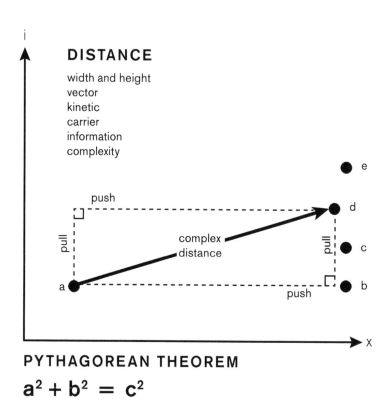

DISTANCE

width and height
vector
kinetic
carrier
information
complexity

PYTHAGOREAN THEOREM

$$a^2 + b^2 = c^2$$

All types of construction, including one's process of thinking, are based on manipulating lines and angles. A vector is a linear distance and its root word means 'carrier' because it carries two values of force: *pull* and *push*. Now consider this, in order to construct something, one must first act like a miner and *pull* on something and then act like a craftsman and *push* it into place. The actions of every mason mirror the form and function of the Pythagorean theorem, and likewise, each of our minds pull and push patterns into place across *imaginary planes*. While doing so, minds forge *abstract* distances in mindspace.

YIN YANG

A balance

The goal of thought theory is to measure the values within symbols. It posits that symbols must be real forms of energy that displace people, since people appear to endlessly chase them. But are people really *chasing* symbols, or are they being *pulled* by them? How do we define the pulling and pushing forces? I believe that it is indeed possible to describe symbolism in terms of energy, spacetime, and gravity. The answer to the unification of matter and symbolism rests with the philosophy of the circle and I believe that a meta-theory explaining symbols with respect to the physics of matter and motion will be revealed through it.

ARCHITECTURE

Explaining the basic framework of mindspace

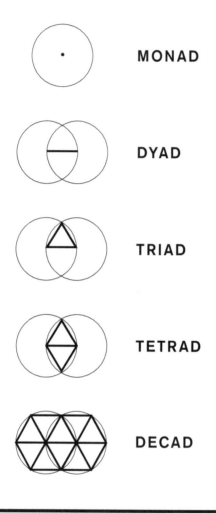

MONAD

DYAD

TRIAD

TETRAD

DECAD

UNION   HIERARCHY

Using reason alone, Pythagoras founded his metaphysical theories upon a set of geometries that interrelate numbers and shapes with meanings. The 5 specific forms above are paramount within thought theory, because they provide the sequence of formation for the **hyperdimensional architecture** that holds mindspace together. The sequence starts with a circle, and expands into increasingly complex geometries. These 'sacred' forms are hypothesized to be the foundation for the residence of minds.

MONAD

Creation

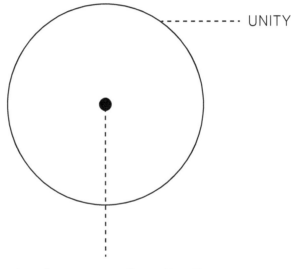

UNITY

center of mass / gravity / attraction

UNION

The base of Pythagorean symbolism is known as the monad (ie. circle). He used the monad as a symbol for God and the Universe. As such, it is the first form. According to Diogenes Laertius, monads evolve into three-dimensional entities, called bodies, that culminate into the four elements of earth, water, fire and air.[8] Does the monad symbolize the singularity found prior to the Big Bang? Does it also symbolize the entire Universe as it expands? How does it relate to the structure of atoms and symbols?

DYAD
Mirror-symmetry

SHARING
Venn Diagram, Vesica Piscis

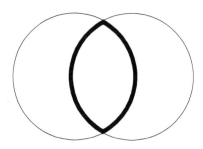

---

folded   unfolded

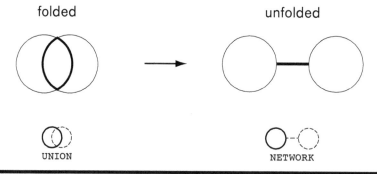

UNION   NETWORK

The word dyad comes from the Greek *dyas* meaning the number two. The dyad can be used to explain a coincidence between any two systems. In psychology, a dyad refers to an interactive pair of persons, such as a patient and doctor, or a mother and her daughter. Its center region, called the *Vesica Piscis*, contains the values shared between the pair. In thought theory, the center region is referred to as a mirror that generates symmetry. The concept of the mirror is important because it is but *similarity* that keeps minds connected in mindspace. For example, a doctor and her patient will interact because 'medicine' is a commonly held value by both individuals. Thus, each of their minds will see values reflected from their shared 'mirror of medicine', and continue engaging one another.

TRIAD

Revealing the depth of a relationship

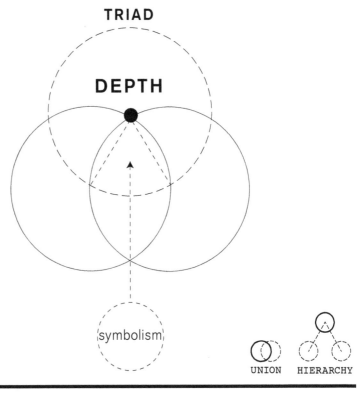

According to Pythagoras and his followers, the number 3 is the noblest of all digits. In thought theory, the number three represents depth, the triad, and it is the 3rd dimension of spacetime (and mindspace). In thought theory, it's given a peculiar role in generating symbolism since it holds the mirror that fits between every dyad. It creates our basic process of comparison and gives rise to mathematical equations; and it's responsible for the success of the Pythagorean theorem. A triad always appears when any two systems are joined and its role is to reflect distances (symmetries). For a simplistic example, consider a cat and a dog joined within a Union, by the thought of being 'household animals'. Now to determine the symbolic value of a cat vs. a dog, a mind will try to measure each member based on the distances it has from being 'household animals'. If a mind must travel fewer distances to reach the cat, a cat will appear to exert a stronger attractive pull on that mind.

SPACE

Two perspectives of the triad

PHYSICAL SPACE

HEIGHT

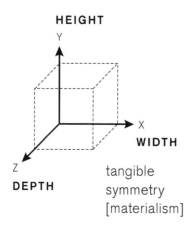

tangible
symmetry
[materialism]

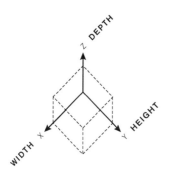

ABSTRACT SPACE

COMPLEXITY

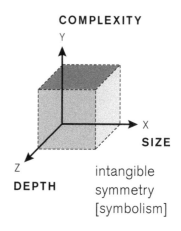

intangible
symmetry
[symbolism]

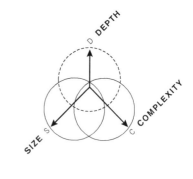

Space is a 3-dimensional framework, because there are exactly 3 planes to travel. According to thought theory, these three planes are created from the triad. Also the triad has two perspectives: physical and abstract. Both perspectives are reflections of the same triad and both function to create 'depth' within mind and material worlds. Thought theory unites both perspectives into a single mindspace containing real and imaginary planes of structure that transcend matter, thought and symbols.

THEORY OF THOUGHT

TETRAD

Mirror symmetry of space

TETRAD

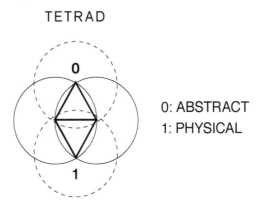

0: ABSTRACT
1: PHYSICAL

NATURAL BALANCED SYMMETRY

mds
model

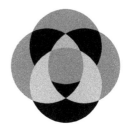

face of
life

UNION

The **tetrad** explains why each group of dimensions - physical, wave, and intelligence - contains exactly 4 dimensions. These arrangements aren't random, but come about as an effect of mirror-symmetry. In metaphysics, the tetrad, and the number 4, are stable forms that represent the natural rest-state for ordered systems. For example, a typical home has 4 corners, a moving vehicle has 4 wheels, and a sturdy chair has 4 legs - anything less is considered unstable. Can the tetrad be the reason why animal faces look the way they do on average? Almost all animals and complex forms of life have remarkably similar facial physiology. Through the eyes of thought theory, stable structures appear to be designed according to the rest-state configuration of a framework of mindspace dimensions.

Forming structure

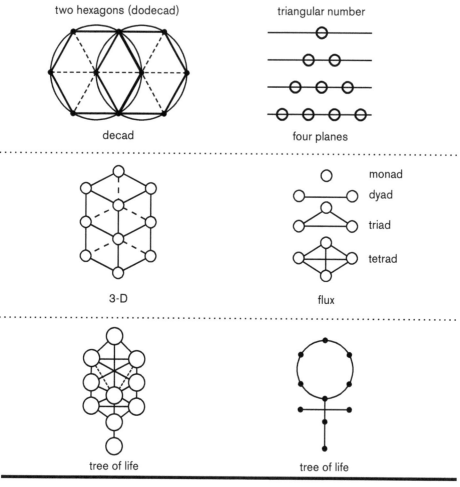

two hexagons (dodecad)

decad

triangular number

four planes

3-D

monad
dyad
triad
tetrad

flux

tree of life

tree of life

According to Pythagorean philosophy, the first four numbers symbolize the harmony of the spheres and the cosmos. They represent the states of flux that build things from unstable to stable states. Notice that the decad, a symbol of the number 10, can be closely related to the dodecad, a symbol for the number 12. These two numbers, 10 and 12, are very closely related in geometry, and this strong relation between the decimal and duodecimal number systems are prominently used within mathematics (ie. geometry). Since mathematics calculates the growth and change of values within any system, the decads above are intrinsic concepts of a system.

TETRACTYS

Elemental model, Pythagoras, circa 500 B.C.

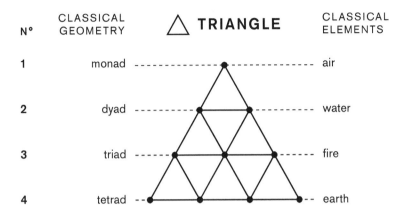

N°	CLASSICAL GEOMETRY	△ TRIANGLE	CLASSICAL ELEMENTS
1	monad		air
2	dyad		water
3	triad		fire
4	tetrad		earth

CLASSICAL NUMEROLOGY

$1 + 2 + 3 + 4 = \mathbf{10}$
$1 \times 2 \times 3 \times 4 = \mathbf{24}$

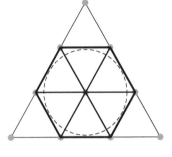

⬡ **HEXAGON**

HIERARCHY

Pythagoras founded the Tetractys to worship the number 10, the triangle, and its other symbolic coincidences. What are all the shapes and meanings that this symbol contains? I will outline coincidences between the Tetractys and increasingly modern schools of thought over the next few pages. Understanding co-incidence is of paramount importance as deeply symbolic ideas overlap and interlock in mindspace across large periods of time.

HEXAGON

Metatron's Cube and the Platonic Solids

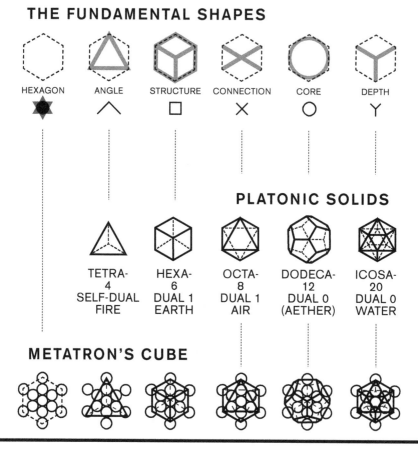

THE FUNDAMENTAL SHAPES

HEXAGON	ANGLE	STRUCTURE	CONNECTION	CORE	DEPTH

PLATONIC SOLIDS

TETRA- 4 SELF-DUAL FIRE	HEXA- 6 DUAL 1 EARTH	OCTA- 8 DUAL 1 AIR	DODECA- 12 DUAL 0 (AETHER)	ICOSA- 20 DUAL 0 WATER

METATRON'S CUBE

The Platonic solids were documented long-before microscopes proved that molecular structures were naturally symmetrical to these patterns. I believe that it is possible to have coincidences between shapes, structures, concepts, properties, numbers, and even letters, on a natural level. For instance, the number 8 could be considered to have some connection to oxygen, because oxygen is the 8th element in the periodic table of elements. Furthermore, *the octagon* has 8 sides, therefore the octagon might also mirror oxygen in some ways. Do the number of these symmetries in nature approach infinity? What patterns emerge when evaluating the relationships of all possible symbols? Is there a single, relational model that emerges which is more useful than others?

THEORY OF THOUGHT

FLOWER OF LIFE

Leonardo Da Vinci, circa 1500 AD

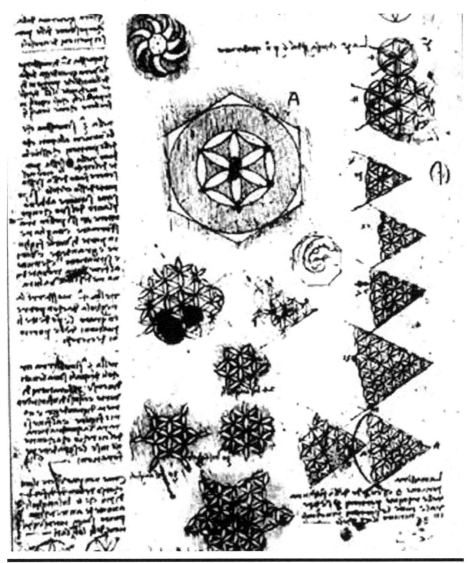

History's greatest minds have deeply pondered the symbolic connections between shapes, meanings, mathematics, and nature. Sir Isaac Newton believed in alchemy. Galileo obsessed over finding symbols in deep space. A strongly rooted Kabbalah symbol called the flower of life must have appeared interesting to them and this symbol still appears in nature hundreds of years later - as the star-shape over the number 8 on an English keyboard.

QUANTUM THEORY

Quantum Chromodynamics, Murray Gell-Man, 1963 AD

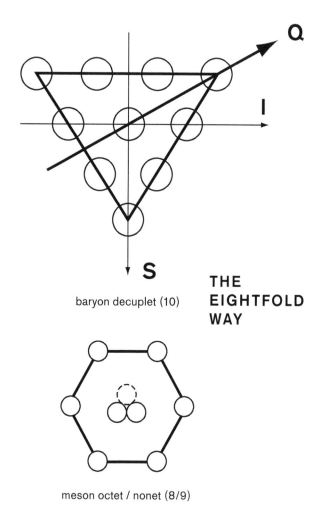

baryon decuplet (10)

THE EIGHTFOLD WAY

meson octet / nonet (8/9)

Murray Gell-Man won a Nobel prize for what he called the 'Eightfold Way'. He purposefully referenced Dharma-Buddha's 'Eightfold Path' in creating a description for quarks. Quarks are the subatomic particles that make up protons and neutrons. It appears that Gell-Man needed to use the Tetractys and the hexagon to diagram quarks because they are essential constructs of the Universe.

THEORY OF EVERYTHING

E8 particle model, Garrett Lisi, 2007 AD

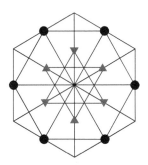

Garrett Lisi captivated scientific minds with his 'Exceptionally Simple Theory of Everything'. Lisi's theory predicts that the standard model of physics should be expanded to 248 possible particles.[9] His E8 diagrams are rooted in the triangle and hexagon, perhaps as an extension of Gell-man's Eightfold Way. It seems that people's ideas are naturally drawn to these shapes because of their simplicity in illustrating nature.

THEORY OF THOUGHT

Metaphysical space, 2012 A.D.

What is a theory of everything? Well, we've believed that our reality has been shaped by symbolism since the dawn of man. However, modern science has done a good job teaching us that reality is quite mathematical and far from whimsical. Thus, a theory of everything must explain the *mathematics* of symbols. Sure, our environments might be seen as atoms of matter moving chaotically in empty space, but this matter is arranged in such a way that it expresses symbolism to any mind that senses it. A theory of everything is an explanation that the material arrangements around us are created by symbolic structures within a hyperdimensional framework of space. However, the question remains, how will this theory of everything help improve the everyday lives of people?

For thousands of years people have spoken of the value contained within symbols, idols, and logos. However this value has always been believed to be mostly unquantifiable. Recently, for instance, marketers have made strides in quantifying the value of corporate symbolism, but I believe there is a more accurate method for calculating abstract value. It can be done for any type of symbol and the calculation will finally uncover the impact its value has on its inter-connected minds. It is hypothesized that symbols behave in way that can be defined, and they interact with other symbols across an imaginary space within a calculable framework of space and time.

Note that each symbol extends across twelve dimensions: time, height, width, depth, meanings, shapes, numbers, letters, period, amplitude, wavelength, and frequency. These are the dimensions of the mindspace - the framework of space that symbols inhabit. Here, symbols enter relationships with other symbols and they interact. As a result of their interaction, the visible arrangement of the physical world is manifested, from the mindspace. When a symbol is highly valued in an environment of the mindspace, it is more likely to be manifested as an arrangement of matter and its value should be correlated to the number of symmetries it shares with the minds within a particular environment. When the value of a symbol fades, its arrangements of matter will fade away with it.

Modelling the interaction of symbols using a theory of everything can lead to many societal benefits. Creating a moving map of symbols in abstract space will allow us to better analyze the motion of people and objects in physical space. It will naturally increase our ability to predict the positive and negative effects that emerge from the events between symbols. We can have a deeper understanding of why particular arrangements of matter are appearing in this part of the world, but not in another. It can lead to better solutions to our material problems by identifying the best positioned minds for moving patterns in mindspace. The right minds might then be motivated to reflect similar arrangements of matter more universally. This theory is peculiarly remarkable because it provides a particle theory that joins mindspace and spacetime into a single meta theory. And if this is true, that it can provide a particle theory along with a theory on minds that control the particles, it will inherently reveal to us the elegant blueprint for the underlying philosophy of the Universe.

If there is a God,
he's a great mathematician

— Paul Dirac

PHILOSOPHY
CHAPTER IV

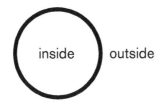 inside outside

DUALITY
&
TRIAD

INSIDE	BORDER	OUTSIDE
symmetry	**attraction**	**negation**
1	2	3

I've gathered these common definitions for the word **dimension**:[1]

1. A measure of spatial extent, especially width, height, or length.
2. Extent or magnitude; scope. E.g. A problem of alarming dimensions.
3. A property. A fundamental measure of a physical quantity.

Thought theory explains that thoughts and symbols are dimensional structures, yet a thought does not seem to have height or width in any conventional sense. Therefore, thoughts must reside somewhat outside of our strict notion of a physical space, and exist within a different type of space - an abstract one that is accessed by neurons. I am sure that the motion of matter is a function of the interactions happening within this hyperspace, and that mathematics unites the abstract and physical realms. So by exploring this idea of mindspace, and the symbolic structures that inhabit it, I'm hoping to uncover specific mind-matter equations.

Our physical space has 4 conventional dimensions to explain the structure of matter. I am adding another 8 dimensions to fully explain the properties of thoughts and symbols and I am merging these 12 dimensions into a framework that I call the mindspace. The mindspace is inhabited by circles and lines, that represent objects, energy, thought, patterns, and symbols. And all these patterns of circles and lines are connected to varying degrees.

In contrast to this symbolic space, objects in the physical world appear to be separated and clearly dispersed across empty distances of space. The physical space is like the inverse of the abstract space, and from a physical perspective it is much more difficult to see the abstract connections that join objects. For example, how do I know if two people are in love if I don't see them together?

To deconstruct the complexity of our shared realities, thought theory extracts science's most important concepts and re-constitutes them under a comprehensive model of relationships. From there, the theory rebuilds a new perspective on materialism, thought, and ultimately symbolism.

Traditionally, humans have been in search of a very important model: the model of the Universe. Models are the starting points for anything we build because they help us define our environments. Scientists create logical models to explore structures and they subsequently draw out diagrams to visualize their models. They join diagrams to create even bigger, more thorough mathematical models. But the real breakthroughs happen when they simplify and unite relatively large and divided models.

On the surface, my model for mindspace is very simple because it shows that the Universe and all of its concepts originate with the circle and are first seeded by its three most basic properties: its inside, its outside, and its border. From its triad of subsections, an abstract space and a physical space emerge that are joined into a hyperspace. Patterns of circles and lines weave this hyperspace together, and an invisible triangle (called the fundamental diagram) governs each pattern according to **three principles**. In the physical space, patterns appear to us as matter and their rules of interaction are revealed to us as 'physics'. In the abstract space, patterns are revealed to us as thoughts, and scientists have yet to precisely define the 'physics' that govern them. Brains, and other biological equivalents, are relationship processing centers that rest between the physical and abstract, while the fundamental mechanism, which manages each invisible triangle is the crucial process that joins thinking and acting. This mechanism combines the three principles of symmetry, attraction, and negation, with those of motion, gravity, and light, and yields the inescapable role of every living being: to function according to these three basic tenants of nature.

The hunt for a universal model on relationships surely began as the earliest of men peered into the night sky. Why? Because the night sky is a looking glass into a vast realm of possibility. What lessons does space contain? Can the star-filled night sky be an echo of a greater wisdom that casts its shadow upon all of our minds?

EMPTY SPACE

Distance and darkness between systems

Space appears to be filled by separated star systems. The stars may seem separated but in fact they are interconnected by light and gravity. Is outer-space the best natural analogy for the human brain and mind?

NETWORKED SPACE

Communication pathways between systems

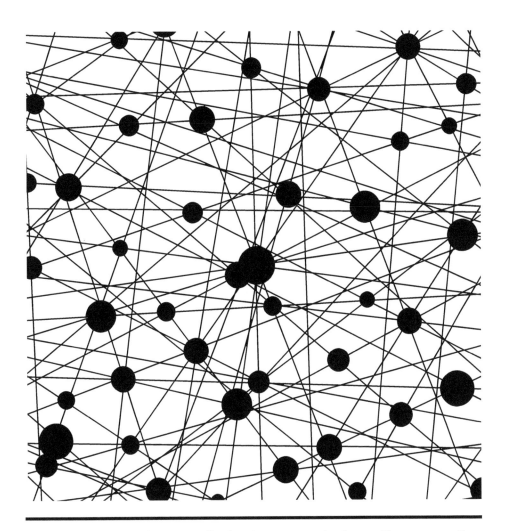

Star systems could potentially be in communication with one another through their light rays. In any graph of systemic communication, a line might represent an impulse between neurons, a chemical bond between atoms, a sound wave between people, cash flow between businesses, a transport route between cities, light rays between suns, or any other pathway exchanging information between organizations.

COMPLEX SPACE

Intelligent arrangements of systems

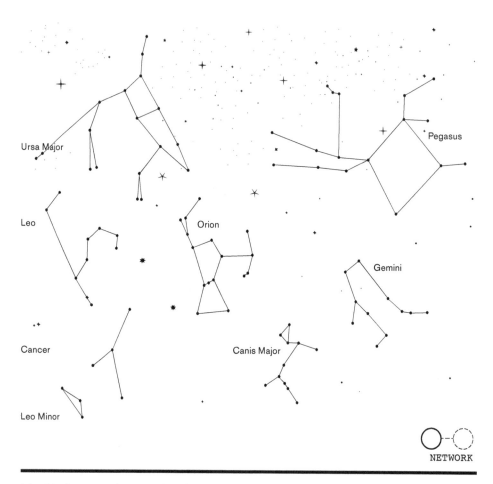

NETWORK

Mankind spent thousands of years ordering the astrological zodiac. Given the vastness of space, it seems that there are infinite possibilities for connecting metaphorical diagrams. But over time, the zodiacal design fell into place in an order of twelve constellations that appear to orbit the Sun approximately every 24,000 years. These patterns must hold significance, and our ancestors must have made specific choices, but what ancient ideas were driving those choices? Can these patterns be identified within each of our unconscious minds?

HIDDEN SPACE

Physical vs. abstract

MATTER　　　　　　　**THOUGHT**

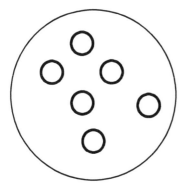 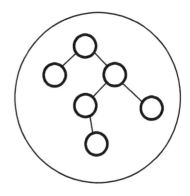

DISCONNECTED SPACE　　CONNECTED SPACE

what we see　　　　　　what we imagine

SYSTEM　NETWORK

There are fundamental differences between physical space and abstract space. In the physical space, systems are distinct entities separated by empty space, but in the abstract space, symbols are all connected and separated by lines. Both spaces are one in the same, however the mind lives in the connected space and the body inhabits the disconnected space. That being said, both spaces work together to help organizations assemble increasingly complex relationships.

ENTIRE UNIVERSE

Infinite scales of patterns

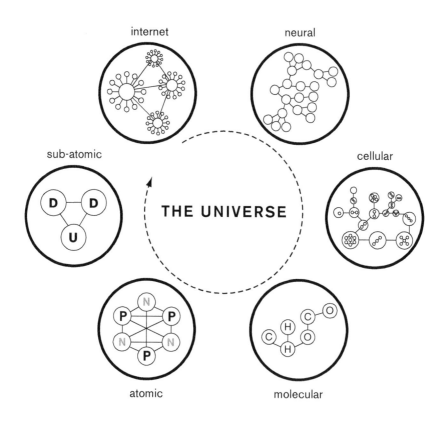

internet

neural

sub-atomic

THE UNIVERSE

cellular

atomic

molecular

SYSTEM   NETWORK   HIERARCHY

Relationships exist at all scales of the Universe, and each one is governed by the same rules. One day, mindspace will be similarly revealed, and an amazing map of symbols will finally been seen.

UNIVERSAL CONCEPTS

Inherent properties of patterns

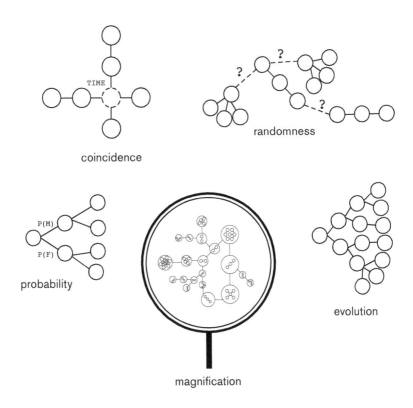

coincidence

randomness

probability

magnification

evolution

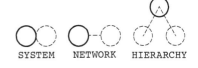

SYSTEM   NETWORK   HIERARCHY

The patterns, that construct symbols, easily illustrate the most basic of concepts such as evolution and probability. These concepts naturally emerge from patterns, and all symbols inherit their properties.

CIRCLE AND THE LINE

Sewing the fabric of reality

O   —

O	—
THOUGHT	ACTION
CORE	PATHWAY
HOME	ROAD
SOMA	AXON
CONCENTRATION	INTERACTION
POTENTIAL	KINETIC
EGG	SPERM
WOMEN	MEN
POSITIVE	NEGATIVE
LOVE	FEAR
ACCEPTED	UNACCEPTED
HEAVEN	HELL
SAFETY	DANGER

NETWORK

All important **dualities** can be easily traced back to the circle and line. Our reality is a fractal, that repeats the concepts of the circle and line into everything that materializes, over and over, *ad infinitum*.

CONCEPTUAL LAYERS

Multi-dimensional concepts

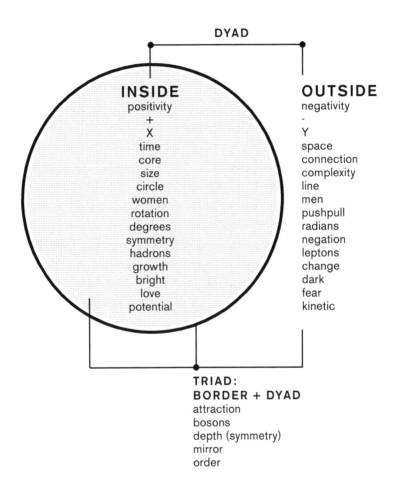

DYAD

INSIDE	OUTSIDE
positivity	negativity
+	-
X	Y
time	space
core	connection
size	complexity
circle	line
women	men
rotation	pushpull
degrees	radians
symmetry	negation
hadrons	leptons
growth	change
bright	dark
love	fear
potential	kinetic

TRIAD:
BORDER + DYAD
attraction
bosons
depth (symmetry)
mirror
order

SYSTEM

Thought theory explains everything by overlaying basic concepts onto its relationships patterns. All organizations are balanced and driven by a complex arrangement of dualities.

HYPERSPACE

Fractal of symbolic time periods

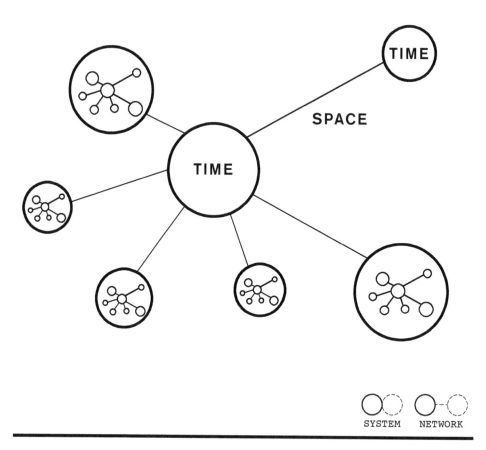

The duality between time and space extends to the basic structure of each mindspace pattern. Time is a potential resource that can be transferred into neighboring organizations by the pathways of space. Therefore, mindspace energy should be considered as a flow of time across space, instead of a flow of space across time. Resources of time ensure the survival of symbols, by contributing towards their **abstract lifespans**. A symbol stores the abstract time within its arrangements of matter, and if a symbol has large resources of abstract time, its material reflections will appear more massive and more successful (important) from outside perspectives.

THEORY OF FORMS

Physical objects reflect from a symbolic world

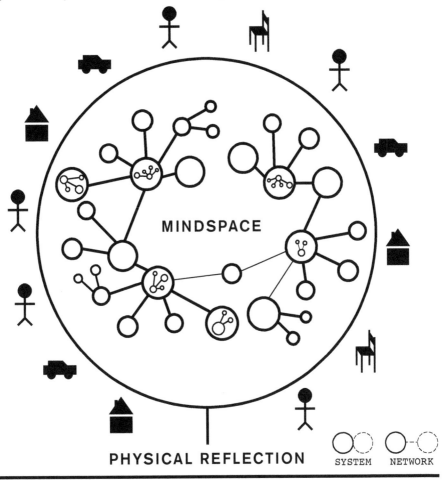

MINDSPACE

PHYSICAL REFLECTION

SYSTEM   NETWORK

Through a metaphysical to physical mechanism, specific material arrangements methodically appear during events in time as reflections of interacting symbols. I will show that this mechanism transmutes a collective
consciousness into a physical reality while adhering to the Copenhagen
interpretation of quantum mechanics.

Plato was an ancient Greek philosopher who lived between 429 BC and 347 BC. He was a follower of Pythagorean ideologies, a student of Socrates, a writer of philosophical dialogues, a mathematician, and the founder of the *Akademia* in Athens. *Akademia* was the first institution of higher learning in the Western world. Plato, along with his mentor Socrates, and his student Aristotle, deeply contributed to establishing the foundations of Western philosophy and science. However, modern scientists have mostly ignored that Socrates and Plato advocated for the existence, and indeed, supremacy of non-physical reality. In our physical world, shadows are a temporary phenomenon produced by physical objects, but according to Socrates and Plato, in their 'Theory of Forms', physical objects and events are themselves shadows of even greater shapes, called Universals, that reside in a hidden world. Essentially, their Theory of Forms describes distinct, yet immaterial substances inhabiting the World of Forms, and our physical world is a shadow of this hidden world.

This book is advocating the same basic ideology, and has reframed the World of Forms into 'mindspace', and the Universal into a 'symbol'. The Theory of Forms was created to address a logical argument for **Universals,** which are abstract qualities or characteristics that are shared between particular things. These characteristics could include being male/female, solid/liquid/gas, or red/blue/green, among any number of meanings. Universals are an ancient problem in metaphysics. From their inception, philosophers have raised questions like: do Universals exist in an individual's brain or in a separate metaphysical domain?[2]

In the 17th century, René Descartes brought Plato's argument further with his theory on Dualism, by claiming that the mind and brain were different entities inhabiting different realms in the universe. Descartes stated that the mind is distinct from the body and its substance is the essence of thought.

These theories, along with thought theory, purport that the nonphysical world can emit, reflect, or purge matter into the physical world. Until now however, there has never been a clear, mathematical explanation for the transfer of energy, or information, between the nonphysical and physical, between thought and matter.

THEORY OF THOUGHT

ANGLE OF DEPTH

Connecting systems into the mindspace

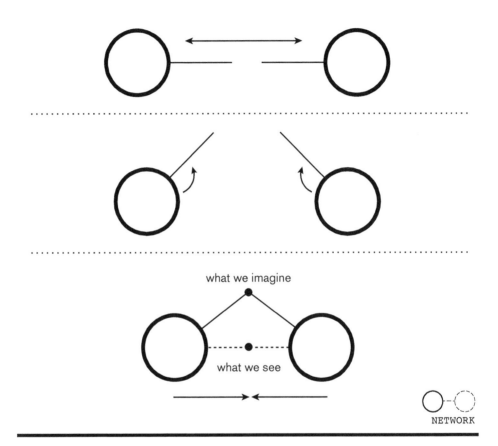

what we imagine

what we see

NETWORK

It's easy to understand that energy can be exchanged between two containers of resources through a straight line. However, the straight-line diagram of the network pattern isn't illustrating the mirror (triad) and its exchange mechanism within the line. It fails to explain that the pathway has a meaningful triangle that governs how the energy travels across it. The method to illustrate the rules of energy transfer is to imagine that mindspace symbols connect to each other across the reflective angle of a mirror. As a result, an abstract triangle rooted in Pythagorean symbology emerges between the containers that is called the fundamental diagram.

DUALISM
Expression of depth

PHYSICAL CONNECTIVITY
'straight-line distance'

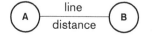

BRAIN

MINDSPACE CONNECTIVITY
'complex distance'

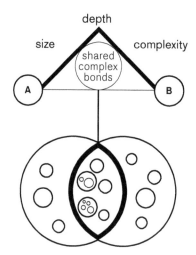

MIND

UNION   SYSTEM   NETWORK   HIERARCHY

Every relationship crosses into depth - an important dimension intersecting both the physical and abstract. The emerging angle represents a complex distance (depth) between symbols and it values their likeness. The rules governing subsequent exchange are governed by the angle and within it, mind, matter, and time are unified.

THREE PRINCIPLES

Principles of a relationship

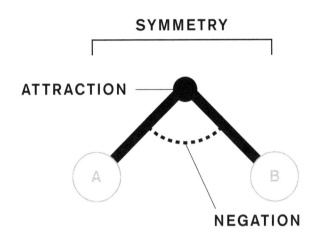

SYMMETRY

ATTRACTION

A   B

NEGATION

PRINCIPLE 1
symmetry
inside
physical
ϕ : 1.618..

PRINCIPLE 2
negation
outside
wave
e : 2.718..

PRINCIPLE 3
attraction
border
intelligence
π : 3.141...

HIERARCHY

Every relationship is governed by an invisible angle ruled by the **three principles** of mindspace which originate from the three subsections of the circle: the inside, outside, and border. All minds access this Λ shape to assemble patterns, because it is a symbol of construction. Note that this **fundamental diagram** is the simplest constructed shape in the universe. As such, any physical road or pathway that forms between two arrangements of matter is first organized in the mindspace across an angle to establish common meanings and purposes. Furthermore, the three most important irrational constants in the Universe are bound to this mechanism by way of basic symbology.

THE ANGLE

Symbolic construction

measurement / construction
compass and square

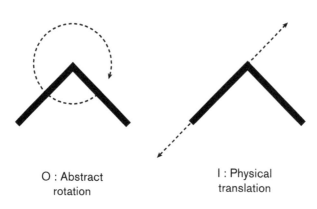

O : Abstract
rotation

I : Physical
translation

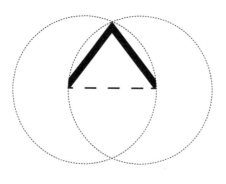

UNION   HIERARCHY

The angle is an incredibly important physical construct and leads to all forms of intelligent design. Each angle in mindspace contains patterns that interact because of mirror-symmetry. For example, when two people speak together, they discuss the same topic - their dialogue is a sequence of mirror-symmetries that keep them connected and engaged.

REFLECTION

Mindspace exists across angles

LENS / DEPTH / INTELLIGENCE / SPINOR

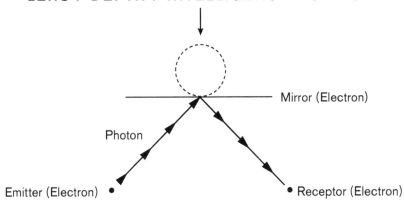

Mirror (Electron)

Photon

Emitter (Electron)

Receptor (Electron)

Simplest notion of reflection between two systems

Reflection is a basic feature of reality because it employs two vectors and one rotation. In quantum electrodynamics, reflection is explained using exactly 3 basic rules concerning electrons and photons. Essentially, this is how reflection works: (1) a photon is emitted from an electron which travels across space until it coincides with another electron. (2) The electron absorbs the photon. (3) A short time afterwards, and at a new angle, the electron emits the photon. These three simple steps are outlined in the diagram above. However, according to quantum mechanics, photons don't really travel across space. They don't even exist until they coincide with another electron. Only a probability of their existence is considered, called a probability amplitude. Photons leap across space and when they are in the middle of a jump, they are under the influence of probability. In thought theory, they are considered to jump into an abstract space, whose higher-dimensional properties are those of *waves* and *intelligence*. This abstraction interweaves quantum mechanics with mindspace, and through complex angles that enter and exit the abstract dimensions, symbols transmit energy to other symbols, reflecting physical reality in the process.

SUPERPOSITION

Abstract space of physical reality

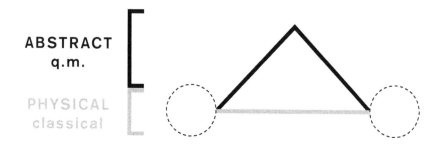

ABSTRACT
q.m.

PHYSICAL
classical

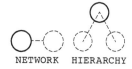

NETWORK   HIERARCHY

Quantum mechanics is a branch of physics that allows for a concept called superposition. Superposition was derived from Schrödinger's equation, and according to the Copenhagen interpretation for superposition, a particle such as a photon, can be in many positions at the same time. This results in an unintuitive and probabilistic view of our Universe, but it might allow for an abstract space of symbols to exist within the superpositioned states of matter. The Copenhagen interpretation calls for a mind to observe or measure an event in order to collapse a particle's wavefunction and bring about its definitive physical state. In thought theory, this process of collapse is defined as the **fundamental mechanism** and it takes place within the angle of the fundamental diagram. The fundamental mechanism acts through the mind and manipulates energy while the mind moves between destinations. I believe that the interaction between our minds across superpositioned states of the Universe is responsible for revealing the world we experience.

QUANTUM MECHANICS

Complex extension of the relationship

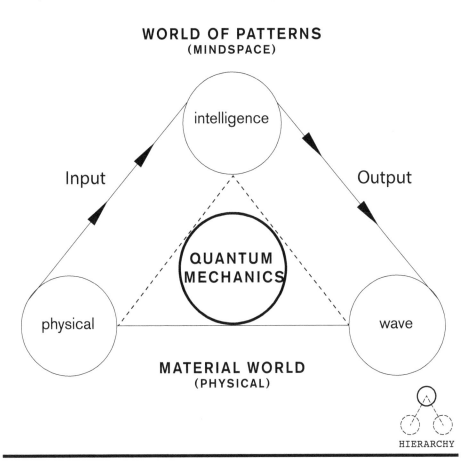

WORLD OF PATTERNS
(MINDSPACE)

intelligence

Input

Output

QUANTUM
MECHANICS

physical

wave

MATERIAL WORLD
(PHYSICAL)

HIERARCHY

A special cycle propagates energy between abstract and physical states. The cycle organizes time into discrete organizations of symbolic instances that share superposition with a number of particles. These superpositioned organizations of time and matter exist to sustain the defined structure of reality. Basically, more patterns equals more matter and at every moment in time, minds collapse patterns and force changes in the distribution of material arrangements across environments. Each cycle of this mechanism also creates a period of time, a physical observation, and an emission of thought. It aligns itself with the Schrödinger's Cat thought experiment and explains that reality is defined only during observation and measurement.

FUNDA-MENTAL

Foundation of the universal fabric

FUNDAMENTAL DIAGRAM
(advanced/folded)

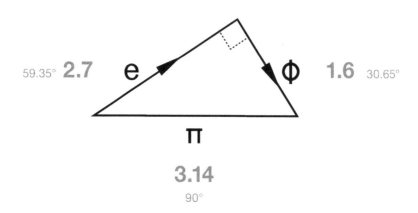

59.35° **2.7** e   Φ **1.6** 30.65°

π

3.14

90°

$1.6^2 + 2.7^2 \approx 3.14^2$

Truncated values reveal a right triangle

16   **27**   **157 x2**

Square Cube Prime

HIERARCHY

This diagram is the more precise version of the fundamental mechanism, and is based on irrational constants. Its shape is formed by a truncated folding process, which will be explained later. The most striking quality of this structure is the relation between π, Φ, and e within a right triangle based on the Pythagorean theorem. It also closely aligns with the 30°-60°-90° sacred triangle. It's even closer to it when the side are π, π/2, and e. The interworkings of these irrational constants create the growing and changing universe that we inhabit. This structure will show how each mind rests on a foundation constructed by irrational constants and sacred geometry.

Ref. page(s): 69
THEORY OF THOUGHT

According to the most accepted theories in physics, time only flows forward*. It may be theorized to possibly flow backwards, however I believe that it is not possible for people to travel backwards in time. It is nearly a certainty that a time machine will never travel into the physical past. What's possible however, is to travel forward into the physical future. According to Einstein's theory of general relativity, it is scientifically possible to build a time machine that can force one forward in time, because of the principle of time dilation. Time dilation states that objects accelerating at different speeds will travel across time at different rates. It is a proven feature of the Universe. The common example describing time dilation is called the 'Twin Paradox'. Twins can age at different rates if one of them accelerates into deep space using a spaceship. When the travelling twin returns to Earth, he finds that the other twin has aged dramatically. He may even arrive far into the future of civilization, when his twin is long deceased. Note that to successfully dilate time, one must produce acceleration using a form of engine.

Theory of Thought reasons that symbols use the fundamental mechanism to affect their acceleration in mindspace and undergo an abstract form of time dilation. This effect can be divided into two ideas: 1) Time has an abstract component and 2) All symbols are essentially time travellers that travel across time at different rates. Let's quickly explore these two ideas.

What is abstract time? We tend to generalize that time ticks forward at the same rate for two people on Earth. However, in abstraction, time does not tick at the same rate. Abstract time flows across the mindspace according to mental activity. What I'm suggesting is that time can be travelled asynchronously across an abstract space, which is embedded into the brain, mind and mindspace. For example, let us consider twins, who with age develop the same goal of becoming the president of their nation. In their lifetime, one person becomes president, and appears successful, while the other does not. One twin had to travel further in the same amount of time to reach her goal. Travel across mindspace does not have to do with the amount of times the Earth cycles around the Sun. Instead it has to do with the amount of cycles that are generated in a mind towards reaching a destination.

*According to QED, only anti-particles can be considered to travel backwards in time. But since our bodies are not made of anti-particles, we can't travel backwards in time.

The more cycles that are performed, across the right abstract organizations, the more likely someone will be forced into a future in which their goals are successfully materialized. In this future, they will appear to have done more, over the same amount of physical time. If we were all given an unlimited number of periods of time, we could all reach extraordinary goals. However, this is not the case and only those of us who travel the mindspace quickly and effectively enough, will appear to reach those goals within the lifetime that we are afforded.

This brings me to my second idea: the only way to speed up across the mindspace and force the progress of abstract time is through the use of symbols. Symbols enable us to complete more cycles faster to reach further pathways into abstraction. Every person must travel the same distance in the mindspace to reach the same goal. Therefore, successful people don't travel less space, they travel the same amount of space, but faster. Symbols provide an additional force that enables interconnected people to reach their goals faster. The abstract acceleration derived from symbols is based on abstract relationships stored within the five mindspace patterns. Abstract relationships, measured in symmetries, provide the fuel for an engine that propels a mind across mindspace.

Symbols push people across time at different speeds, which results in the difference between personal success at one end of the spectrum, and some form of extinction at the other. The effect applies to living people as well as all other arrangements of matter. As the material forms of a symbol are forced into an abstract future through time dilation, its material arrangements will appear to gain strength and power and will appear to out-succeed other material arrangements within its environment. Ultimately, time dilation is the base mechanism for the evolution of species and it is also the cause for the extinction of all non-relatable life forms and objects.

Theory of Thought is an exploration of the time machines that permeate our environments. These hidden, mechanical engines (built on truncated geometries) force minds and groups of minds into living extraordinary lives and are governed by a single set of meta-physics founded on modern science.

*One of the greatest synthesis
of the 19th century was the
synthesis of the laws
of electricity and magnetism,
with the laws of the behavior
of light, so there was not two
things but only one*

— Richard Feynman

ATTRACTION
CHAPTER V

ATTRACTION

GRAVITATION
positive

ELECTROMAGNETISM
negative

ORDER

Driven by the principle of attraction

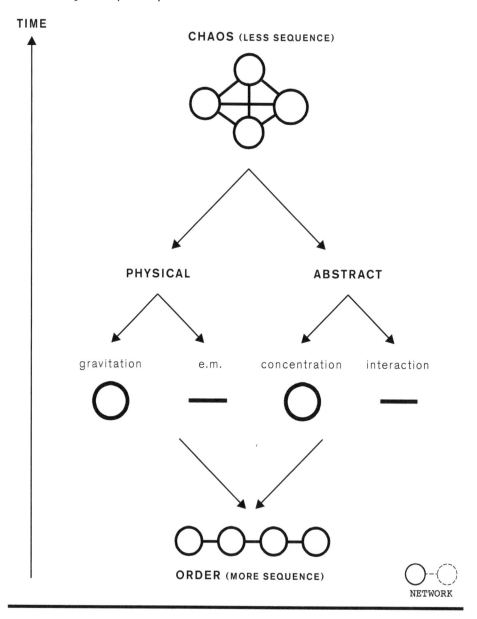

TIME

CHAOS (LESS SEQUENCE)

PHYSICAL

ABSTRACT

gravitation e.m. concentration interaction

ORDER (MORE SEQUENCE)

NETWORK

All living beings are conduits for a force of nature called attraction. Attraction is an assembly process forcing an order between circles and lines within mind-space, and it combats against the perceived flow of time, entropy and disorder.

In thought theory, energy is contained by the circles and lines of mindspace. **Chaos** is the natural state of energy, meaning that without a process of **order** the circles and lines become randomly dispersed. Thought theory's principle of Attraction explains how dis-ordered patterns are re-arranged naturally in mindspace. Attraction's relational pulling force can be defined in a physical and abstract manner. In the physical space, attraction manifests as **gravity** and **electromagnetism** between objects, but in the abstract space, it manifests as **concentration** and **interaction** between symbols. Notice in the next two paragraphs, how I simply compare the notions of our physical nature with those of the abstract space:

Let us start with the *physical space*. Gravity is the force of attraction exerted by the size of an object. It is a force that pulls smaller organizations into larger organizations, so it mainly works between organizations of *different scales*. For example, gravity is exerted between you and the Earth and it feels like an inescapable pulling force. Electromagnetism, on the other hand, is emitted as a line and it works between organizations of similar scale. For example, electromagnetism keeps a pen between your fingers. Your fingers, the part of your body that interacts with the pen, is on a similar scale than the pen. It's important to realize that electromagnetism feels like *manipulating* matter while gravitation feels like *absorption*.

Let us move onto the *abstract space*. Concentration is the force of attraction that is bound to the circle. It is a force that pulls smaller organizations into larger organizations. So it works between organizations of different scales. For example, concentration is exerted between you and the Earth and it feels like 'attention' and admiration. Concentration creates awareness. The other abstract force of attraction is called interaction. It is bound to the line and it works between organizations of similar scale. For example, interaction works between a pen and your fingers. Although your mind may concentrate on a subject to write about, interaction causes your fingers to move a pen while thinking about the subject. Its important to realize that interaction feels complex, while concentration feels simple.

Overall, attraction brings people, objects, ideas and events together and it also crosses into decision-making and behavioral patterns. Any two things that move closer together are under the influence of attraction, whether it's planets, people, or letters on a page. The physical and abstract notions of attraction must be considered when calculating the motion of matter, because people and objects are displaced by forces of attraction exerted on a physical and intellectual level, at the same time.

There is another mindspace principle called **Negation** that mitigates Attraction and gives systems the ability to separate. Negation emerges from the four dimensions of waves and opposes Attraction in many philosophical respects. It is a fundamental principle that comes about from the 'outside region of the circle', and results in **destructive complexity**.

Here's a common generalization about gravity: objects with the greatest size exert the greatest amounts of gravity. Consider that the largest mountain on the planet, containing the most atoms and mass, is Mt. Everest. I have proposed that gravity and concentration are the same force, and that concentration can be understood as attention or awareness. Therefore, Mt. Everest must generate the most amount of awareness since its the biggest mountain. So what is awareness? Consider the thousands of people who discuss Mt. Everest every day. When people think about a relatively broad subject, that's awareness, and it's part of the concentrative process. But does Everest *attract the most visitors* because of its concentrative force? It turns out that Mt. Fuji is the world's most visited mountain, and it's much smaller than Everest. So what's really forcing people to move towards Mt. Fuji, since it must produce less awareness from its lesser size? The answer is that Mt. Fuji contains a higher complexity than Everest. Fuji has ski parks, restaurants, and other 'attractions', that spring from highly connected arrangements of patterns. It may not be the most physically massive mountain, but it is the most intelligently organized. Its intelligent arrangements provide it with a higher interactive component that provide it with 'abstract' forms of mass. Its abstract mass exerts additional units of attraction that pulls on the bodies of people. So a large physical mass will cause people to become attentive and aware, but a large abstract mass forces people to approach it physically. I believe that these concepts can be applied to everything we think about and engage with.

ATTRACTION

Process of order

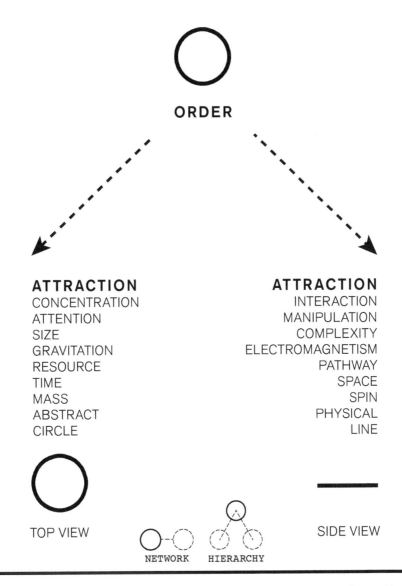

ORDER

ATTRACTION
CONCENTRATION
ATTENTION
SIZE
GRAVITATION
RESOURCE
TIME
MASS
ABSTRACT
CIRCLE

ATTRACTION
INTERACTION
MANIPULATION
COMPLEXITY
ELECTROMAGNETISM
PATHWAY
SPACE
SPIN
PHYSICAL
LINE

TOP VIEW

NETWORK   HIERARCHY

SIDE VIEW

Both physical and abstract forms of attraction originate from the two perspectives of the circle: its top view and side view. And both perspectives are necessary to explain how every unit of symbolic matter moves through the hyperdimensional mindspace.

MOLECULES

Systemic order

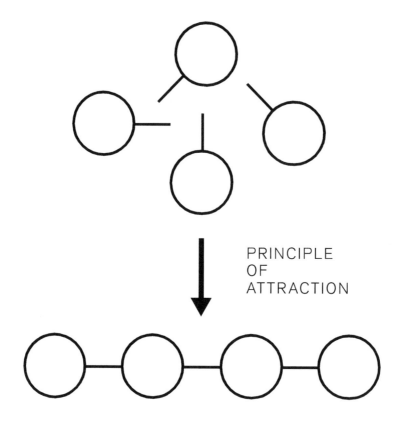

PRINCIPLE
OF
ATTRACTION

Assemble into stacks

NETWORK

Using the forces of attraction, and against the natural flow of disorder, patterns join into complex arrangements of ordered stacks. When the stacks gather enough size, they physically manifest as molecules, amino acids, DNA, organs, and higher forms of complex structure. The physical manifestation of patterns from mindspace results in the arrangement of matter across every scale of the Universe.

THEORY OF THOUGHT

MATTER

Bohr's first model of the atom

ATOM OF HYDROGEN

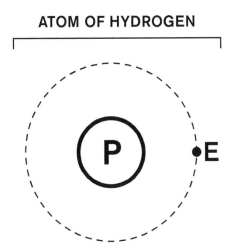

ATOM OF HYDROGEN

GRAVITATION ELECTRO-MAGNETISM

HYDROGEN PATTERN

SIZE COMPLEXITY

NETWORK

The basic structure of an atom can be easily compared to the basic unit of a network pattern. Look closely - the proton, which exerts gravity, and the electron, which exerts electromagnetism, mirror the circle and line of the pattern. This diagram shows how physical matter is actually bound to the structure of abstract patterns.

UNITS OF STRUCTURE

Bridging the abstract and physical into one

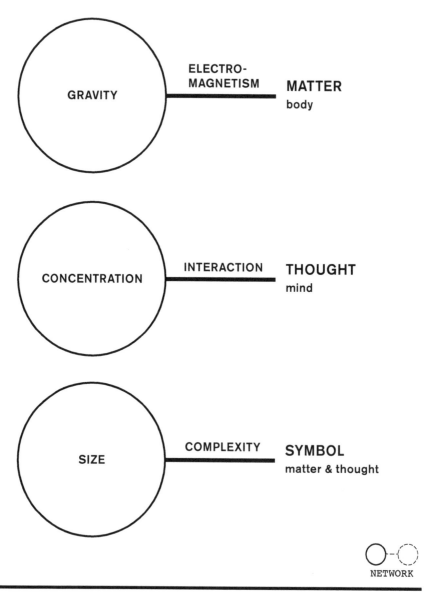

GRAVITY

ELECTRO-
MAGNETISM

MATTER
body

CONCENTRATION

INTERACTION

THOUGHT
mind

SIZE

COMPLEXITY

SYMBOL
matter & thought

NETWORK

By layering concepts, physics and metaphysics can be joined into a single theory governing one space. Our bodies and minds are both directly connected to a single mindspace by way of the relationship patterns.

Our brains and minds construct patterns, and they have to mentally force these patterns together. There are two different techniques to construct them: 1) concentration (pull) and 2) interaction (push). When one starts concentrating on an idea, broadly related ideas will appear within their region of mindspace. This notion can be interpreted in two ways. First, you can think of mindspace to exist within one's brain, and concentration would then represent a gathering of related thoughts or ideas when broadly thinking. The second way is to think of a mindspace as an environment of physical matter that extends beyond the brain. In this context, concentration gathers arrangements of matter across environments. So by concentrating on a subject, related arrangements of matter, such as books and pictures about the subject, start appearing around you. This book strongly advocates for the latter view on mindspace, although both contexts are correct. The truth, as it seems, is that both views of mindspace (brain vs. environment) affect one another probabilistically because they are interwoven into a single mind-matter paradigm.

A mind first starts working in the mindspace by concentrating. It groups a large number of related ideas into a single 'concentration' - which equates to gathering an organization of patterns. During the process of concentration, thoughts fall into place like the objects that fall toward the Earth from the process of gravity. Once concentrated, a mind uses interaction to arrange the patterns. Interaction functions according to the same guiding principles as electromagnetism. For example, electromagnetism is guided by quantum mechanics and wave interference. Therefore, patterns of thoughts can be likened to waves and can interfere with one another. Poorly arranged patterns generate friction and cause frustration in the mind, and people get upset because their patterns are poorly aligned across their regions of mindspace. Upsetting emotions erupt when some patterns fail to align constructively, proceed to generate a form of friction, and go on to destroy the arrangements of other closely grouped mental patterns.

I am hypothesizing that the mind uses abstract versions of gravity and electromagnetism to order itself. And as a result, all of the effects of gravity and electromagnetism, on time and space, are also effects of the mind. For example, in thought theory, all symbols can warp the fabric of mindspace, experience abstract time dilation, and create friction; to name only a few of their metaphysical properties.

GRAVITATION

Concentrative type of attraction

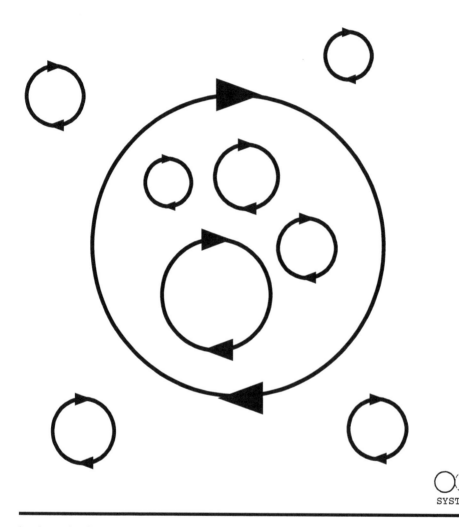

SYSTEM

In thought theory, gravity and concentration are the same concept, and their purpose is to increase the size of a circle by grouping in smaller, nearby circles. For example, a brain concentrates on a subject by grouping several patterns together. Thought theory hypothesizes that the process must begin with a single pattern at the top of a hierarchy, that proceeds to travel down its branches to combine several related ideas.

CONCENTRATION

Focus on size

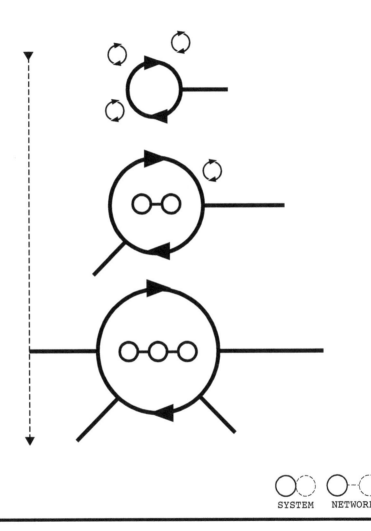

SYSTEM   NETWORK

Concentration and interaction go hand in hand. As an organization concentrates, additional pathways form between nearby systems, thereby increasing complexity and the potential for interaction. As a mind assembles its symbols into increasingly large groups, the resulting symbols will have a greater number of pathways leading to other minds and symbols.

ELECTROMAGNETISM

Interactive type of attraction

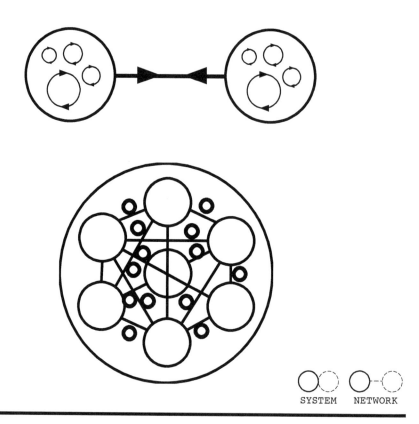

SYSTEM   NETWORK

When one pattern is too large to concentrate into another, the resulting activity is much better described using the rules of electromagnetism (interaction). Electromagnetism is an interaction between electrons, photons and nuclei. In contrast to gravity's generalized absorption force, electromagnetism is a targeted force, like a hand grasping a pen. In the context of physics, interaction and electromagnetism are based on the rules of quantum mechanics, and unlike concentration and gravity, interaction is an exchange between different stacks in separated hierarchies. Note that the exchange process between symbols is governed by interaction, and thus electromagnetism (e.m.).

INTERACTION

Focus on complexity

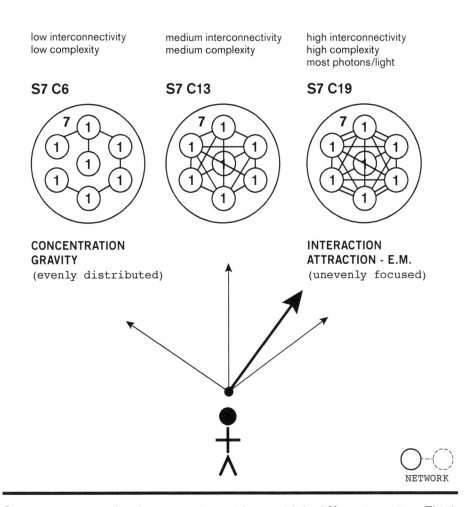

low interconnectivity
low complexity

medium interconnectivity
medium complexity

high interconnectivity
high complexity
most photons/light

S7 C6

S7 C13

S7 C19

CONCENTRATION
GRAVITY
(evenly distributed)

INTERACTION
ATTRACTION - E.M.
(unevenly focused)

NETWORK

Organizations can be the same size yet have widely different c-ratios. Think of a rock and a pen of very similar masses. A rock is incredibly simple, while a pen is much more complex, in part because a pen contains some forms of glyphs. While concentration might draw someone to both the rock and pen equally, complexity forces them to take hold of the pen instead of the rock, because the mind is strongly attracted to complex organizations and their wider range of functions. So when distance is equal, people are drawn to objects of complexity over simplicity.

ORBIT

Bodies cycle and fall towards a center

**SOLAR
SYSTEM**

*Not to scale

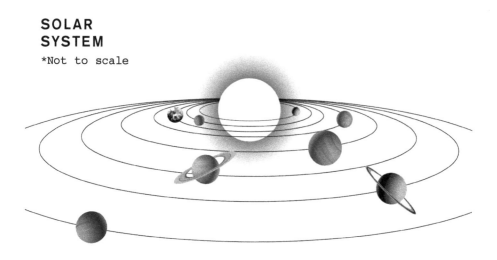

The orbit is an inescapable feature of space. Objects in deep space orbit one another due to inertia and gravity, and are subject to friction. I would argue that the Sun and its interaction with nearby bodies is a beautiful symbol of attraction, in plain sight. Similarly, minds place patterns into periodic orbits to limit their collisions in mindspace, and this natural thinking process projects every symbol into a circle, a cycle, and a period.

Have you ever believed that some object held some weight in your mind? Unlike other comparable objects, this special object contained historical, beautiful, or perhaps spiritual value to you. I believe that this feeling of weight actually exists and should be translated into a concept called abstract mass. This weight comes about from relationships - so any mind that consciously perceives these relationships will feel that weight. So to clarify, there is a weight that results from *the number of atoms*, and there is another type of weight that results from *the number of relationships* (between atoms). Let me provide an example about the effects of the weight of a pen to someone who cares deeply about pens:

People usually won't travel great distances to get to a pen. When people do travel large distances for tiny objects like a pen, it's because there are a large numbers of relationships forcing them to do so. For example, a person might own a pen store. A store is a collection of patterns bound together into a large, symbolic organization. This organization might even include other minds, such as workers, customers, bankers, admirers, and so forth. A collection of minds is bound by complex relationships and their complexity generates strong attractive forces that warp regions of mindspace and create accelerated areas. This is exactly why a pen store owner can travel the globe for a small pen, while a normal person quite simply cannot. Large collections of symbols shift the execution speed of events by displacing more information across more pathways in the same amount of time. So if one's mindspace is warped by large groups of symbols, reaching some goal will take a much shorter amount of time. A **warp** in a mindspace can extend into far physical distances and a person's decision to travel the world for any single object is caused entirely by the attractive forces exerted by his mind's proximity to these warps. Decision-making patterns are indicative of the warps people travel across mindspace and decisions are by-products of travelling minds that are inherently pulled by nearby symbols. By the way, education is the primary means to discover these warps and structure new warps, which in-turn produces behavioral variations. Therefore it is much more probable to see a pen store owner travel across the globe for a pen, than a person that does not have the relationships to motivate such behavior.

PHYSICAL MASS

Material substance

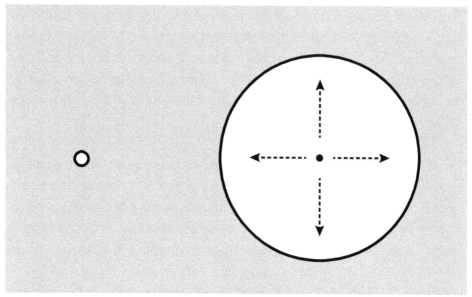

Less size
Less awareness

More size
More awareness

Which perimeter do people notice first?
Which perimeter do people spend more time looking at?

SYSTEM

We need to keep track of the largest objects because they have the greatest potential effect on our lives. As such, we tend to know what the tallest building in the world is, while often ignoring the smallest building. In thought theory, there is a clear connection between time, awareness, and size. People spend a lot more of their time studying larger objects and organizations. This behavior happens because larger objects exert more attraction, and we're more or less subjected to them. Remember that all symbolic objects survive off the time that our minds supply them. So we feed these large symbols by giving them our time while they pull on our minds. As each symbol *acquires* more time, it will even appear to grow in material size, but as a symbol *spends* its time (by way of entropy) its associated matter will appear to fade away. So each symbol must continually feed on our minds to survive the flow of disorder. It's also very important to consider that the displacement of a symbol can possibly force the displacement of other connected minds and bodies.

ABSTRACT MASS

Hidden value

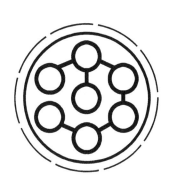

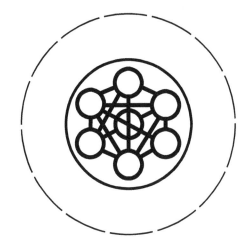

LESS LINES

LESS ABSTRACT MASS

MORE LINES

MORE ABSTRACT MASS

NETWORK

In thought theory, physical mass is correlated to the sum of atoms, while an **abstract mass** is correlated with the sum of lines between atoms. Together, physical size and abstract size (complexity) combine to form the *total mass* of a symbol in mindspace. Total mass can be quickly depicted by a larger, secondary perimeter made from the total sum of lines. This second perimeter allows us to visualize the additional range of the interactive force. The expanded perimeter allows for a larger pool of symbols to exist within a single environment, and these symbols will appear to exchange greater amounts of time. This secondary perimeter is a warped region of mindspace that forces material displacement while speeding up the execution of events.

Furthermore, in thought theory it is hypothesized that Dark Matter is a function of abstract mass (and the concept of lines). This will be discussed in a later, increasingly scientific edition of Theory of Thought.

INTELLIGENCE
Knowledge of the abstract dimensions

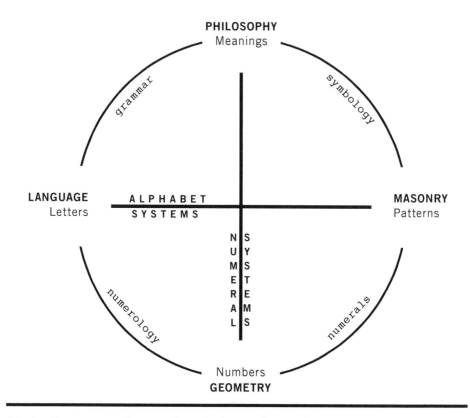

PHILOSOPHY
Meanings

grammar

symbology

LANGUAGE
Letters

ALPHABET
SYSTEMS

MASONRY
Patterns

N S
U Y
M S
E T
R E
A M
L S

numerology

numerals

Numbers
GEOMETRY

Abstract mass can be simply called complexity. Complexity, as a mass, is a measure of symbolic information and hierarchy. Look at it this way, the Sun has an incredibly large physical perimeter. Therefore it must also have a large perimeter in mindspace. But consider that a Sun is composed primarily of hydrogen and helium atoms bumping around each other, which equates to a relatively low abstract mass. In contrast, a human body has a very small physical perimeter. It does comprise however of enormous numbers of purposeful arrangements. It contains complex patterns of molecules, cells, veins, and organs, etc. Its inner-relationships can be summed and equated as additional values of complexity. All things contain hidden complexity and when a mind is well educated in alphabet and number systems, it discovers and builds upon these hierarchies of information. Uncovering and assembling hierarchies increases abstract mass, builds warps, leads to a form of abstract time dilation that brings about future events faster, and creates the perception of *'success'*.

CYCLING

An ending flows into a new beginning

PUSHPULL + ROTATION

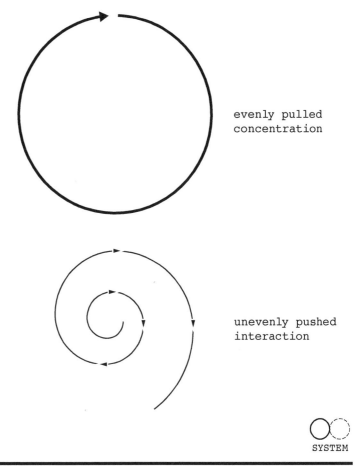

evenly pulled
concentration

unevenly pushed
interaction

SYSTEM

The first diagram above explains a system with a neutral, or equilibrated c-ratio. Any increase or decrease in the c-ratio leads to the second diagram. Complexity produces a rotational friction that affects the diametric rotation of organizations in mindspace. Organizational diameters are dynamic entities and abstract mass manipulates them while forcing the production of spirals throughout mindspace. Furthermore, the spiral represents the warp that changes the probabilities of events occurring between symbols.

PERIODICITY

Cyclical behavior of a mind

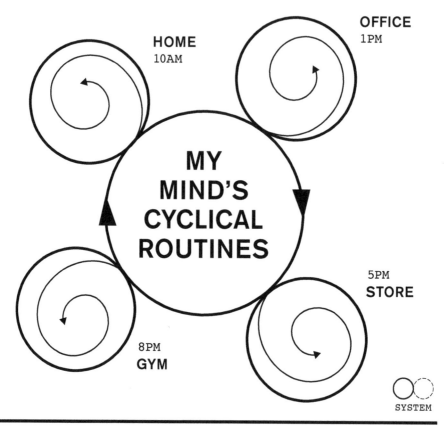

All human behavior is governed by **periodicity**. Everyday we tend to repeat the same patterns as the previous day. It's just like a circle, but we call our form of periodicity 'routine'. We generally acknowledge that routine is healthy for the mind. Within the periods of routine, like that of eating dinner, exist even smaller periods, such as taking dishes out, using them, and putting them back. Our minds spiral into and out of symbols, forcing our bodies and the objects around them to behave periodically. Periodicity is also the reason that our hearts continuously beat, and our lungs continuously breathe. Spiraling periodicity is a function of the physical, intelligence, and wave dimensions acting together to produce dynamic and sustainable environments for complex arrangements of patterns.

Ref. page(s): 95 **THEORY OF THOUGHT**

MOTIVATION

Pulling and pushing force between minds

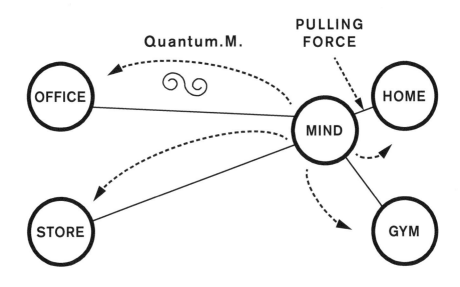

PULLING
FORCE

Quantum.M.

OFFICE

HOME

MIND

STORE

GYM

NETWORK

Symmetry is a measure of common **environments** and it exists within the abstract lines between symbols. Anytime the word 'like' can be employed to describe two symbols, some amount of symmetry can be considered to exist. For example, a cat is *like* a dog, in that they are both household animals. When the connecting line of symmetry is drawn short, there are less differences between two symbols. The strength of attraction between symbols is dependent on the lengths of these lines. Furthermore, as one approaches a symbol, they are transferring time into it. This exchange of time produces a shift in environment, which is associated with abstract time dilation, and it creates the perception of material displacement along with the execution of events.

MOTION & PROBABILITY

Different approaches on attraction

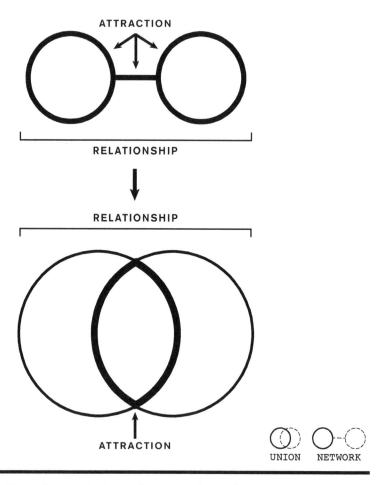

ATTRACTION

RELATIONSHIP

RELATIONSHIP

ATTRACTION

UNION   NETWORK

The network pattern demonstrates a linear pathway between symbols and the pathway symbolizes the transfer of smaller-symbols, particles, or any type of information between environments. But by using the union pattern, we can understand that anything transferred between symbols, is in fact shared by both symbols. The network pattern and the union pattern are different interpretations of any relationship - the network pattern describes motion, while the union pattern describes probability.

ENVIRONMENTS
The intersection in a union pattern

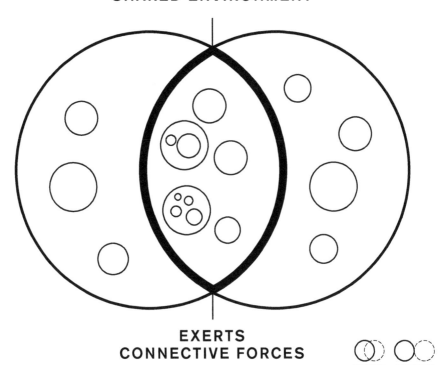

SYMMETRIES
SHARED ENVIRONMENT

EXERTS
CONNECTIVE FORCES

UNION   SYSTEM

A person's probability of interacting with another organization is calculated using the union pattern. If two organizations, such as a person and a car, share objects, such as a car key, drivers license, auto insurance, gasoline, and so forth, then they will be more likely to interact together. These commonalities are various symmetries shared between organizations and they appear in the **Vesica Piscis**, the center region of the union pattern. The more symmetries that are found in this center region, the more likely two organizations will interact and be found in closer proximity. *For reference: an environment is like a photograph and a symmetry can be any object that appears within the photograph.*

I'm going to provide an example of attraction using a house and a car. The measure of symmetry between the two objects is found in its shared environments. For example, a house usually has a driveway. A driveway is where a car parks. Driveways are a common environment that are shared by both houses and cars. Houses can also have garages. A garage is another common environment. Houses are usually attached to roads, structures designed for cars. These common environments create events between cars and houses. If you can photograph a house with a garage, a driveway, and a road, the photo will probably contain a car. The coinciding symmetries will naturally attract cars towards the house. Even cars that drive near the house, but don't belong to the house, are probabilities of attraction. Now imagine a house without a garage or a driveway. What are the probabilities that there will be a car nearby? Subtracting these commonalities lowers the chance any cars will be photographed near the house. However, a car can still be connected to the street adjacent to the house. Now remove the street. Envision a house in the middle of nowhere surrounded by trees (since there are no roads). In this situation there is even less chance that one would picture a car nearby. This example illustrates that it is only commonality (ie. relationships) that dictates attraction, interaction and the probability of objects coinciding in mindspace.

PROBABILITY OF OBSERVING A CAR AND HOUSE WITHIN THE SAME SPACE

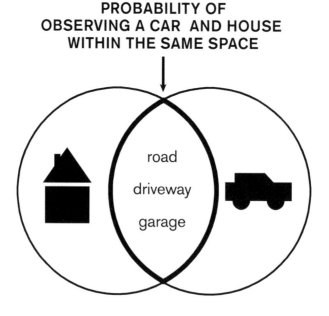

road

driveway

garage

UNION

Mindspace is a framework of symmetry and measuring symmetry reveals a distance between mindspace symbols. Symmetry leads to attraction and it is a hierarchy of similarities between a car and a house, a man and his photo, or a corporate culture and its logo. Symmetry is often hidden within intelligent concepts, yet it is always connecting binary opposites, exact copies, and everything in between. It is the source of the attractive forces that bind all mindspace symbols. By evaluating the degree of symmetry between things, one can develop a model of the forces that produce events. Relationships with higher degrees of symmetry will have higher levels of interaction *vis-a-vis* relationships with lesser degrees of symmetry. Furthermore, 'order' forces symmetrical symbols closer together in the mindspace and as a result, symmetrical symbols will appear closer together in physical space. Order is a categorizing force that makes sure that objects are kept organized. For example, in forests, trees grow near other trees, while in a supermarket, products are organized near similar products, such as milk being placed next to cheese. This practical ordering process is a natural function of intelligent behavior. It might appear that humans are making decisions, however there are no decisions, only forces of attraction that order sequences of symbolic events in the Universe based on relationships.

Is randomness, or otherwise known as equilibrium, a real phenomenon? If the state of a system is considered random, is it possible to be made more random? Surely not, since it has already achieved a state of randomness. In contrast, if a system's state is perfectly ordered, can it be made more ordered? The answer is yes it can always be increasingly ordered, because reaching a state of perfect order is impossible. Reaching a perfect order is akin to the idea of reaching infinity and it is embodied in the notion of the circle and π. Decisions always move towards order, as a function of attraction, and thus making random decisions is impossible. All imagined possibilities emerge from randomness and are fueled by attraction's desire to produce order. Therefore randomness is a state, not an objective or a phenomenon that helps reach objectives. Randomness is really no different than the concepts of ignorance and indifference as they all provide the lowest level denominator for explaining the arrangement of intelligent structures. Remember that order is the living force tasked with combating the basic flow of randomness in the Universe.

All of physics is either impossible or trivial. It is impossible until you understand it, and then it becomes trivial.

— Ernest Rutherford

PHYSICS
CHAPTER VI

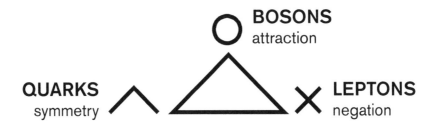

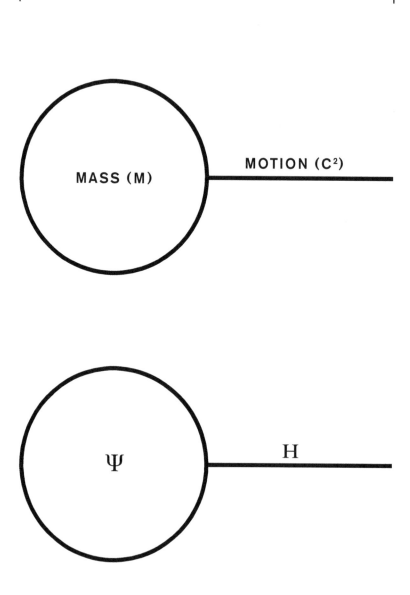

Einstein's special relativity introduced physicists to the mass-energy equivalence principle, otherwise known as $E=mc^2$. It explains that **energy** and **mass** are united into a single concept. Energy is the abstract half of the concept because it is only *indirectly* observed in nature as a quantity of possible work. Mass, on the other hand, is its physical half because it can be directly measured as the weight of moving matter. As different sides of the same coin, energy is considered to be a property of mass, and mass is considered to be a property of energy.

Note that the mass of an object is a function of its velocity/momentum, which is why Einstein relied heavily on the concept of 'rest' mass (ie. moving at the same speed and direction).

Since total energy is described by both the circle and line, the circle represents rest mass, while the line represents the velocity component of the mass (the line's maximum value is that of lightspeed). Furthermore, consider that every object has a value of physical energy, while also having a value of intellectual energy. Every material book, building, or other object contains a form of mass that is held within its visible glyphs and only a mind can evaluate.

From a traditional physics perspective, 'a rock' and 'a bottle of Coke', of the same rest mass contain roughly the same amount of energy. However, from the perspective of a mindspace, a bottle of Coke will contain more energy than a simple rock of similar rest mass. The additional energy exists because a bottle of coke contains specific arrangements of relationships - ie. designs, words, numbers, letters, ingredients, etc. Since these arrangements provide some additional mass, they must be represented by circles - that exist as abstract values to be exchanged between bodies of rest mass. Therefore each line in mindspace can be also visualized as a (dotted) circle of abstract mass, like so:

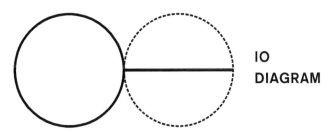

IO
DIAGRAM

As I've outlined previously, the role of the linear pathways of mindspace is to exchange rotating circles (of time) according to some rules of mindspace that arise from a hidden triangle. I believe that scientists have called this process quantum mechanics and the invisible triangle has been named the wavefunction. In this chapter I will introduce the underlying philosophy of quantum mechanics, where it comes from, and its role in governing the superpositioned symbols shared by all living minds in mindspace.

You may be thinking - am I talking about physical reality, or some imaginary region that exists in our minds? - and the answer is both. Quantum mechanics and symbolism can be integrated into a single framework that governs everything, from what we see, to what we think - if physics and psychology are re-explained with respect to the 5 relationship patterns.

If you don't already know, quantum mechanics is a non-classical set of rules that govern the tiny particles of matter that interact together across the physical world we see around us. But unlike classical mechanics, the rules of q.m. describe fuzzy and unintuitive behaviors. For instance, a baseball (ie. classical) does not behave like a particle (ie. non-classical) even though a baseball is made entirely of particles. There is a different set of rules that govern the universe on the tiniest of scales, and to understand its philosophy with respect to the human mind is to reveal a theory of everything. Thought theory takes the approach that these tiny particles are visible fragments of symbols and they exist with regard to the non-classical manner in which our minds function.

Philosophically speaking, quantum mechanics cannot exist without minds, because quantum mechanics comes about from the act of measuring the motion of stuff across the lines of mindspace. I believe it can all be generally explained quite simply - a circle will rotate across a line, in effect creating the line - and as it rotates the Pythagorean theorem reveals a triangle (fundamental diagram) which is imbued with the three principles of mindspace. This triangle gives rise to our notion of the wavefunction, and when it's measured by a mind, it emits different types of particles into reality (depending on its form). The fundamental mechanism explains what arrangements of matter and how much of them will *probably* appear at specific positions within any given environment. Generally, the mechanism I'm describing is designed to explain *very large groups* of protons, neutrons, electrons and photons. The exact arrangement of those particles, will reveal classical objects, with positions that depend on the arrangement of symbolic patterns (which includes the apparatus' that scientists use to measure such objects).

Overall, this chapter should reveal a scientific philosophy expounding that the laws of physics, matter, and all other organizations, tangible and intangible, are based on the guiding principles governing the architecture of patterns within mindspace. This chapter will also show how all material economics, and specifically particle physics, originate from a universal cycle. I believe that understanding cycles of patterns will provide the foundation of an *actual* theory of everything which explains the connection between the physics of spacetime with those of the human mind.

**THOUGHT
THEORY**

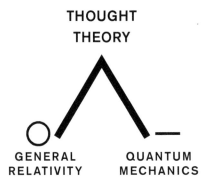

**GENERAL
RELATIVITY**

**QUANTUM
MECHANICS**

**IS
THEORY
OF EVERYTHING**

Currently, there are some physicists that are actively pursuing a theory of everything, but can they provide us with this theory? A theory of everything should explain human interaction, it should explain love, ideas, and success. Physicists might someday build a complete particle theory, but is that everything? I believe that physicists are interested in a theory of everything because their field does indeed hold the keys to understanding ourselves, and consequently everything. But I can't identify an existing branch of physics that could potentially weave its concepts with those on the human mind. This procedure might be better accomplished by an inter-related branch, such as metaphysics. For example, it is still unclear to physicists why there are exactly three families of particles and how this fact might descend from some primordial language. Theory of Thought provides the breadth of reasoning that emerging metaphysicists will need understand this primordial language, and in contrast to a traditional book on physics, Theory of Thought reveals *why* 3 particle families exist: each family represents one of the *three* basic properties of the circle - the inside, outside, and border.

THE STANDARD MODEL

Modern laws of physics

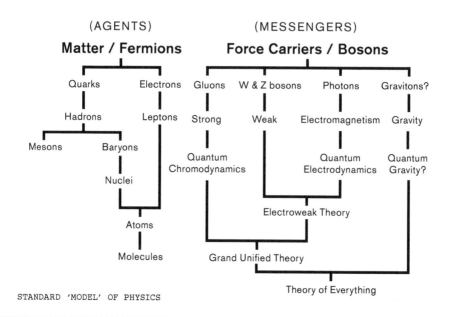

(AGENTS)

Matter / Fermions

(MESSENGERS)

Force Carriers / Bosons

STANDARD 'MODEL' OF PHYSICS

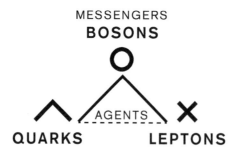

MESSENGERS
BOSONS

AGENTS

QUARKS   **LEPTONS**

HIERARCHY

The Standard Model is the most significant mathematical model in physics. It describes a collection of particles that emerge from the respective force fields of nature. However, its contextual description lacks any philosophical base and raison d'être. In contrast, thought theory's fundamental diagram provides an underlying philosophical structure revealing a hierarchical language that creates every unit of 'depth'. The Λ, O, X shapes are fundamental shapes of mirror-symmetry and they provide additional clues for revealing the basic architecture of hyperdimensional space.

PARTICLES

Re-visualized using patterns

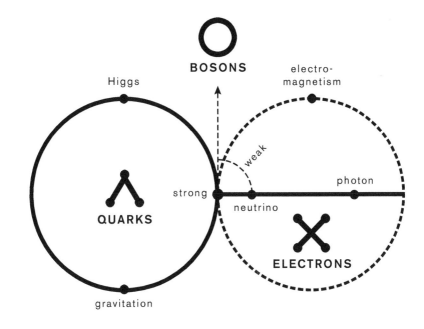

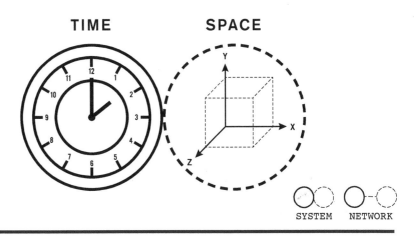

The entire Standard Model can be transposed over thought theory's IO diagram, revealing a model for the general assembly of particle fields. Every type of particle represents a junction in a diagram that is hypothesized to be the simplistic foundation of particle physics.

MODELS

Illustrating depth

IO DIAGRAM

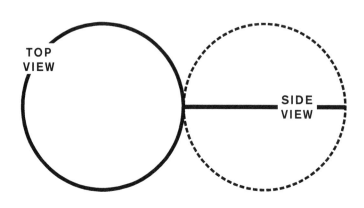

TOP
VIEW

SIDE
VIEW

FUNDAMENTAL DIAGRAM

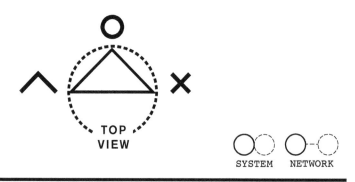

TOP
VIEW

SYSTEM   NETWORK

Each pattern exists within two views: time (top view), and space (side view). The views exist because the concept of orthogonality. The side view appears when the top view is rotated 90°. While the circle rotates (represented by the dotted circle), a concept of 'depth' emerges and is embodied within the line. The basic structure of 'depth' is a form of mirror-symmetry and it can be generally called the fundamental diagram. As such, the laws of quantum mechanics are based on orthogonal symmetry, and while the FM diagram explains the formation of particle matter, it also explains the formation of symbolic patterns in mindspace.

Ref. page(s): 52, 152 **THEORY OF THOUGHT**

BRIDGING

Mindspace is an angle away

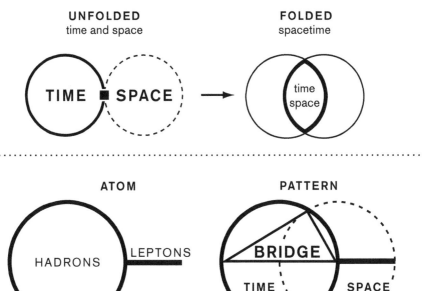

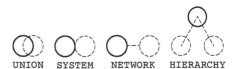

Time and space represent the inside and outside of a circle, respectively, as shown in the IO diagram. The inside circle (top view) is the origin of symmetry, and the outside circle (side view) is a form of depth held within by an abstract wave (Λ) created by rotation. The fundamental mechanism starts when the inside and outside are merged into a union diagram in a process called **folding**. Their fusion reveals an abstract mathematical bridge that organizes units of depth in mindspace. It acts as a needle and thread that weaves the mindspace framework into the material fabric of spacetime.

FOLDING

Forcing the fabric of spacetime

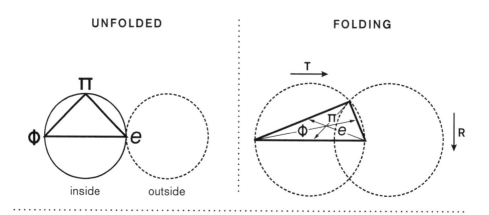

UNFOLDED

inside   outside

FOLDING

FOLDED

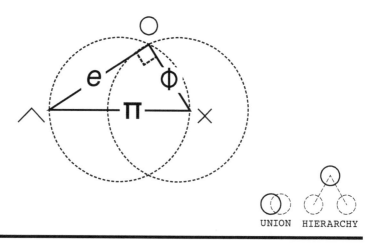

UNION   HIERARCHY

The fundamental diagram has an unfolded (monad) and folded (dyad) state. The folding process is driven by two basic motions, **rotation** and **translation**, that force circles across lines with respect to the Pythagorean theorem. This basic process reveals a union pattern with a special right triangle emerging at the critical juncture between space and time, matter and mind, and rationality and irrationality. *Irrational number truncation will be discussed in the next chapter.*

Ref. page(s): 69   **THEORY OF THOUGHT**

CONTAINMENT

Creating a system in mindspace

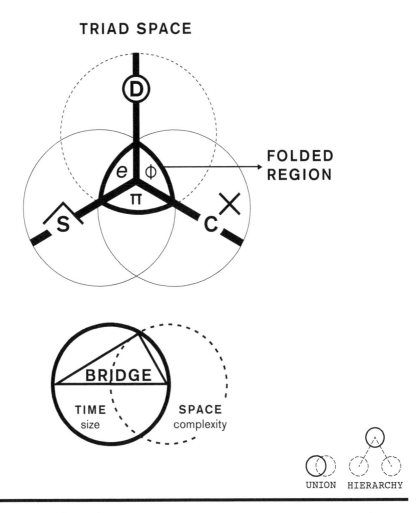

TRIAD SPACE

FOLDED REGION

BRIDGE

TIME
size

SPACE
complexity

UNION HIERARCHY

The folded region (depth) is structured by the symbolic dimensions of intelligence and it is the origin for the abstract pulling forces in mindspace. Through it, depth, size, and complexity are interrelated into three-dimensionality, and intertwined between the physical and the abstract. In short, this region binds subatomic particles with some symbols in mindspace, and is held together by the three most important irrational constants in the Universe.

CHARGE
Providing life to matter

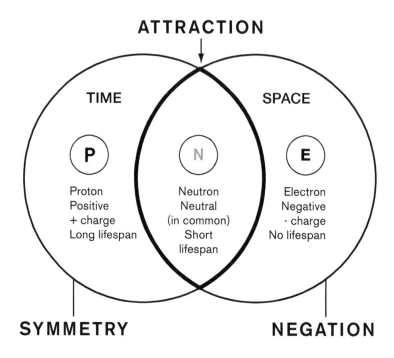

ATTRACTION

TIME

SPACE

P

N

E

Proton	Neutron	Electron
Positive	Neutral	Negative
+ charge	(in common)	- charge
Long lifespan	Short	No lifespan
	lifespan	

SYMMETRY

NEGATION

UNION

UNION

The three most important particles of nature are the proton, neutron, and electron. They can be logically divided across the sections of the union pattern, and each particle is closely related to a particular fundamental principle: symmetry, attraction, and negation. These are the only 3 atomic particles that are found in every object, body, and thing the Universe, and all other particles are either quarks, bosons, neutrinos, or 'exotic' (ie. found at unusually high energies such as the Big Bang, a sun, or a black hole).

THEORY OF THOUGHT

ROOTS OF THE DYAD

Base of hyperdimensional space

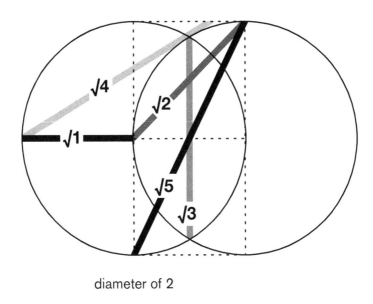

diameter of 2

UNION

The Pythagorean dyad is perhaps the simplest structure containing the most amazing mathematical implications - its structure holds within it the √2, the √3, and the √5. Other than π, Φ, and *e,* they are the three most important irrational constants in mathematics. Considered a sacred geometry, the dyad reveals an eye-like shape that has been accepted as the focal point of numerous religions. It often symbolizes the all-seeing eye of a god. This structure might be well positioned to be the base of the universe because it contains numerous references to mathematical 'roots' (pun intended). I will argue that symbols are being formed in the mindspace atop this irrational framework. Over the next several pages, I will illustrate how this shape can be placed at the center of physics and how these three square roots can be identified at the base of quantum mechanics.

BOSONS

Fundamental forces of attraction

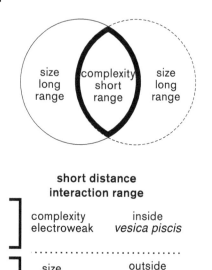

Bosons Physical Forces	Components of Attraction	
Strong	Complexity	
Weak	Complexity	complexity inside
Electromagnetism	Complexity	electroweak *vesica piscis*
Electromagnetism	Size	size outside
Gravitation	Size	relativity *vesica piscis*

short distance interaction range

long distance concentration range

An atom is the smallest structure that exerts all **four physical forces**. These forces are transmitted by a family of particles called **bosons** that form the frame of the dyad to create interaction between hadrons (ie. protons) and leptons (ie. electrons). In thought theory, *hadrons* are considered 'agents' of time, while *leptons* are 'agents' of space, and *bosons* are 'messengers' between the agents of time and space. Hadrons emerge from time, the 'inside' of the circle and *concentration*, while leptons emerge from space, the 'outside' of the circle and *interaction*. Bosons are in-between these two distinct regions.

Every messenger (boson) has an observed particle, except for gravity. No one has ever observed the theoretical boson of gravity called a Graviton, which inhabits the boundary of the circle. It is more likely that gravity is not a particle field, but a warping field or another abstract notion of *concentration* that is released exclusively by the inner agents of time. Note that like gravity, time itself is also an abstract concept lacking tangibility. In contrast to gravity, the other particle bosons are tangible in nature and must thus originate from the agents of space.

THEORY OF THOUGHT

Every boson is an embodiment of attraction and as a result each type of boson is categorized by size or complexity. This subdivision of attraction determines the type of strength exerted by each boson: Bosons of the *size* subdivision, like gravity, are weaker than bosons of the *complexity* subdivision. Bosons of the *complexity* subdivision, including strong, weak, and electromagnetic forces are much stronger. However, complexity's strength is only exerted over *very short distances*. On the other hand, gravity may be weak, but it interacts over *very long distances*. Also, there are features of electromagnetism, such as radiation and electrostatics, which along with gravity, interact weakly over very long distances.

I can make several conclusions based on these ideas. The first is that electromagnetism is divided between short and long range, and therefore must be partly communicated as concentration by the agents of time. The reason that electromagnetism is subject to both time and space, can be understood through the *Vesica Piscis*. The short range forces occur within the *Vesica Piscis*, while the long range forces occur outside of it. And since electromagnetism is coordinating a circle that is moving across both divisions it must maintain a concentrative aspect. Furthermore, it could be said that Quantum mechanics occurs inside the *Vesica Piscis*, while classical mechanics occurs outside it. As a result, the outer region of the *Vesica Piscis* is structured by the inverse-square law, while the inner-region is structured by the complex numbers and irrational constants found in non-classical physics.

Using these ideas, I can also make the case that in everyday life, interaction is a much more powerful attractor than concentration, but over very short ranges. For example, a very large building in front of me may be quite attractive and capture my awareness. However, that networked computer in my pocket is more likely to steal my attention away at any given moment. Computers are masses containing many communicating symbols and as a result of their complexity, they exert strong, attractive forces on nearby minds. Big buildings on the other hand, exert a relatively weaker force and are easier to psychologically disengage. The reason these comparisons are real and not fictitious, is because the physical and abstract spaces share a single space, and the physical forces that we observe in nature are just physical reflections of the equally real, abstract forces that rule over the patterns accessed by our minds.

FOUR FORCES

Abstract nature of physics

Physical force	Abstract force
GRAVITATION	CONCENTRATION
ELECTRO MAGNETISM	COMPLEX INTERACTION
STRONG FORCE	BINDING TO TIME
WEAK FORCE	ROTATION OF SPACE

Nature has created equivalent concepts for the physical and **abstract forces** that permeate mindspace. In effect, there are invisible circles travelling (rotating) across invisible lines, and its all just an abstraction for the motion of time across mindspace. Within this abstraction, the strong and weak forces hold a circle to a line while it rotates across it. In physics, these two forces are responsible for holding an atom together and in thought theory, they explain how to hold the dyad together and enable the creation of the triad. The triad is then represented by the other two abstract forces, concentration and interaction, that forge a depth whereby time is exchanged between symbols.

FORCE FIELDS

Frame of the fundamental mechanism

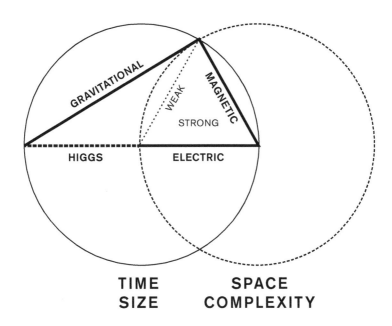

TIME
SIZE

SPACE
COMPLEXITY

UNION

UNION

Time and space are different entities with different functions, yet they are united in a mechanism that manages patterns. The forces of nature are exerted by a framework (fm diagram) that binds the circle (time) and line (space), and each force field performs a specific task that helps build and maintain this pattern in mindspace. Every pattern goes on to reflect at least one atom.

*E.m. is split into short and long ranges because the line is the side view of a circle. Gravity isn't split between ranges, because the circle is **not** the top view of a line (did you get that?). This is the most basic difference I can find between quantum mechanics and general relativity. Also note that the vector of gravity does not cross into the Vesica Piscis.*

Simple division of time and space

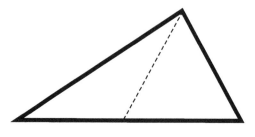

SIZE / TIME (MASS)

RELATIVITY / LORENZ SYMMETRY

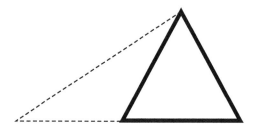

COMPLEXITY / SPACE (MOTION)

ELECTROWEAK / GAUGE SYMMETRY

UNION

Time and space share a common frame, and that's why they can work together seamlessly, despite being ideologically separate entities. Bosons assemble the frame of spacetime because they are the messengers carrying information between vertices. As a feature of their differences, time only has access to the relativistic subdivision, while space only has access to the quantum mechanical subdivision. Both of these subdivisions overlap, but not completely. The following conclusions are once again made by observing the dyad and the *Vesica Piscis* between time and space.

INTEGER SPIN

Magnitude of the angular momentum

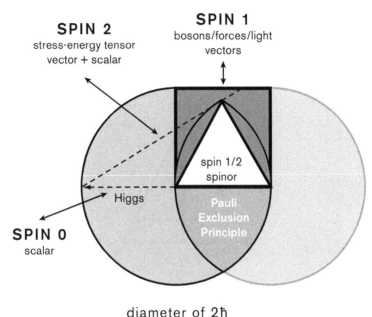

SPIN 2
stress-energy tensor
vector + scalar

SPIN 1
bosons/forces/light
vectors

spin 1/2
spinor

Higgs

SPIN 0
scalar

Pauli
Exclusion
Principle

diameter of 2ℏ

spin 1/2

$$S = ℏ × \frac{\sqrt{3}}{2}$$

spin 1

$$S = ℏ × \sqrt{2}$$

spin 2

$$S = ℏ × \sqrt{4}$$

UNION

Bosons (photons, weak, and strong) have integer spin, while fermions (protons, neutrons, and electrons) have half-integer spin. Integer spin gives a photon the ability to overlap other photons in quantum space. This quantum capability allows bosons to maintain complex connections, as superpositioned networks, bridging **fermions** (agents of time and space). Unlike bosons, fermions abide by the Pauli Exclusion Principle. The Pauli Exclusion Principle appears to be a result of the *Vesica Piscis*, since fermions are generated entirely from within the *Vesica Piscis*, while bosons are not. Furthermore, the magnitude of the spin angular momentum is tied to the irrational values within the dyad, in the reduced Planck scale.

HALF-INTEGER SPIN

Two-valued quantum degree of freedom

SPIN 1/2

720° ROTATION

spinors
baryons/matter/fermions/leptons

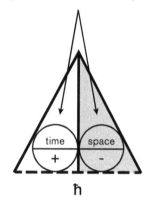

ħ

ħ/2   ħ/2

ROTATION 2	ROTATION 1
360°	360°
reveal particle(s)	assemble particle(s)
time	depth
t	x

ψ0(x, t)
probability density = probability of assembling and revealing a particle
A spinor determine a probability amplitude for the quantum state.

A fermion has a special spin that essentially forces it to rotate twice before returning to its original state, meaning that it must rotate 720° to complete one quantum cycle. Quantum spin can be perceived as motion across a Möbius Strip. It could be said that the dyad functions like a Möbius strip and each 360° rotation of spin interacts with one of two circles of a dyad, therefore two rotations cycle a single dyad. The first rotation acts across space, while the second rotation acts across time, and a fermion particle emerges from the Vesica Piscis once a full cycle of the dyad is completed.

QUARK MODEL

Baryon octet overlaid onto a baryon decuplet

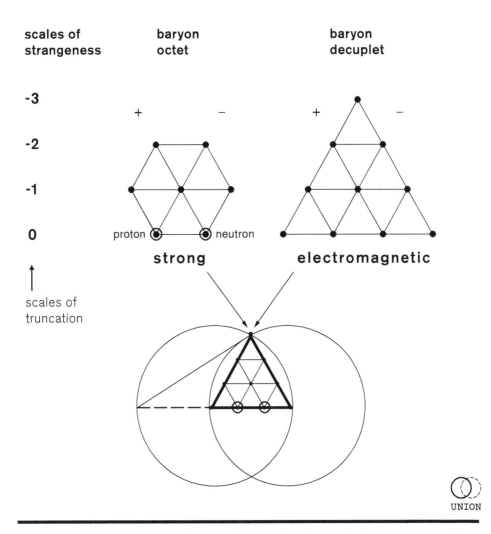

The baryon octet and decuplet can be seamlessly overlaid onto one another. It's remarkable that the Tetractys, worshipped by Pythagoras thousands of years ago, actually represents the structure of matter. It also appears that matter is divided into 4 scales. Baryons with -3 strangeness are considered the heaviest forms of matter, while baryons with 0 strangeness are the lightest forms. It's also important to recognize that the matter of our everyday world is all composed of the lightest scale, perhaps as a result of its specific position on the frame of the fundamental diagram.

BREAKING SYMMETRY

A flow across a hidden mechanism

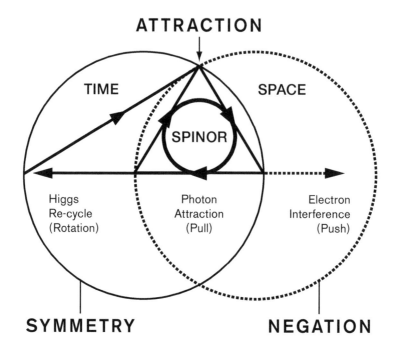

ATTRACTION

TIME   SPACE

SPINOR

Higgs
Re-cycle
(Rotation)

Photon
Attraction
(Pull)

Electron
Interference
(Push)

SYMMETRY   NEGATION

UNION

In the center of the mechanism, particles are assembled using the quantum mechanical spinor. The FM mechanism fuels a spinor across the abstract dimensions of mindspace, and it produces particles that are 'entangled' to symbols in mindspace. Basically, it ejects meaningfully associated particles into our physical world. Also, like the weak interaction stipulates, the mechanism is chiral and asymmetrical - meaning that it prefers to flow in one particular direction (ie. symmetry breaking). As the spinor rotates, and during each periodic cycle, it breaks symmetry and reflects arrangements of matter in physical space. Reality can now be generally understood to be simply driven by an abstract mechanism that manipulates arrangements of symbols and advertently controls the appearance and disappearance of matter within specific environments of spacetime (through rules of probability).

I have deep faith that the principle of the universe will be beautiful and simple

— Albert Einstein

One might ask, what's the point of all these diagrams? What do they mean to me? These diagrams represent the way that particles are distributed around us. These diagrams might one day explain why particular arrangements of matter are present within one environment, yet absent from another. To understand, assume that every person, object, or atom can be described as an organization bound by time and each organization contains a fundamental mechanism:

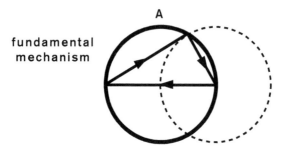

Organizations can be of vastly different scales but they still interact with each other, using their inner-organizations. They share and distribute energy and matter between themselves. So, if you pick up a book from a table, you are displacing matter from one organization and placing it into another. The exchange of matter between organizations is described by stacking the mechanisms:

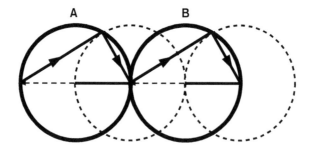

The way this process works is by connecting organizations together and using their mechanisms to order complex arrangements of matter/symbols across mindspace. Basically, each mechanism works according to supply and demand, and its purpose is to establish equilibrium between itself and its neighbors. Nothing in this universe is immune to supply, demand, and equilibrium because they are foundational processes driving the behaviors of everything. In physics, temperature and pressure always attempt to establish equilibrium. In economics, the laws of supply and demand explain the fluctuation of prices towards equilibrium. In thought theory, equilibrium is at the center of the fundamental mechanism and supply and demand are the guiding principles for it as they guide the exchange of energy/matter across a network of hyperdimensional mechanisms. Thought theory's version of the supply and demand mechanism explains the exchange of material arrangements within all environments. It is a mind-matter displacement mechanism using the concepts of growth and change as abstractions for the concepts of supply and demand (and gravity and e.m.).

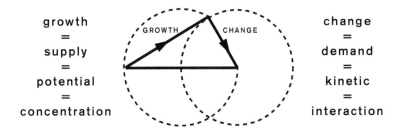

growth	change
=	=
supply	demand
=	=
potential	kinetic
=	=
concentration	interaction

Every organization in mindspace is a mechanism for establishing growth and change, and each mechanism functions by growing potential energy and changing it into kinetic energy. Essentially, all organizations are either growing their stocks of energy into a pool of abstract time, or spending their abstract pool of time to accelerate arrangements of matter across space. Stated slightly differently - every living being is an organization in mindspace that is performing work, and is managed by a **cycle of work** that functions according to the economic principles of environmental supply and demand.

Since our brains manage the growth and change that we undergo, our thoughts must be reflective of the functions of the fundamental mechanisms located at the center of our mind. At some points in time our thoughts appear to be within a growth phase, called concentration, and at others they are in a change phase, called interaction. In everyday practice, all thoughts fluctuate rapidly between the two phases, while completing work cycles in the brain. Over many cycles, a group of thoughts (or those of an entire symbol that emit the thoughts) will reveal an average distribution over the growth and change phases of the work cycles. The average distribution of thought over each phase will directly affect a body's physical activity, concerning its acquisition or dispersion of matter. So if one is thinking about growing, they will attempt to concentrate matter *around* their body. But if one is thinking about changing, they will attempt to move matter *across* their body. This rule of thumb concerning the behavior of a body, applies to the distribution patterns of any person, symbol or organization in general.

GROWTH

Concentration of patterns

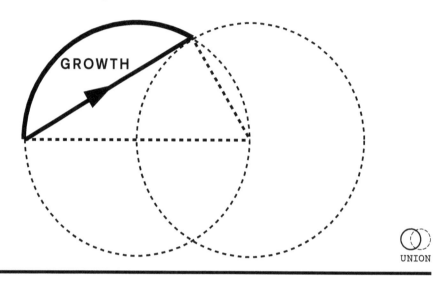

UNION

The cycle of work (ie. fm mechanism) is comprised of 2 main phases, called growth and change, and 1 equilibrating phase called balance.

Let's consider a factory and a consumer: a factory is a supply-oriented organization, while a consumer is a demand-oriented organization. In order to build inventory for consumers, a factory must heavily exploit the growth phase of the mechanism. In doing so, it accumulates matter, or more precisely, it accumulates the possibility of matter. The growth phase represents the first phase of work, which collects patterns in abstraction. If patterns are well-collected, inventory will appear within the factory's environment as the balancing phase is completed. A great way of understanding the process of growth is by comparing it to the idea of work. The reason people work is to *accumulate* a quantity of stuff, and at some point in the future they will *spend* their stuff to induce change in their lives. This initial phase in the cycle of work assembles the potential energy (ie. time) in order to get stuff.

Going a bit further - money is a tangible replacement for growth and it represents the time stored within collections of patterns, which in-turn represents influence over matter. When consumers have money, they have access to a pool of potential material resources, and when they spend money, they trade from that pool to command bits of matter into desired positions.

CHANGE

Interaction of patterns

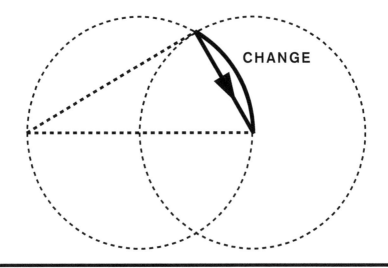

CHANGE

UNION

Every work cycle always comprises of an accumulation and a spending phase. One can never spend more then it accumulates, and will often spend much less, especially once friction is factored in to the cycle. What is being accumulated and spent? - Energy / Resources / Time (all of which are the same).

The phase following growth is called change, and its closely related to kinetic energy. Unlike growth, change is associated with space and physical motion. During a change phase, matter will appear to move, especially since it transpires within the *Vesica Piscis*.

Change causes matter to move *across* an environment and any consumer exploits the change phase by exchanging one set of resources for another. It's the necessary component of 'exchange' and it transfers energy between interconnected cycles of work. It is also associated with the collapsing of the wavefunction, the expenditure of time, and the spending of money.

Growing and changing can simply refer to the two basic functions of life: working and enjoying. These two concepts are forged into our minds by the dyad architectures of mindspace, and living is entirely made possible by the manipulation of dyads using these two geometrical phases.

BALANCE
Equilibrating growth and change

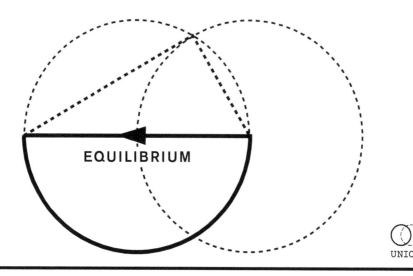

EQUILIBRIUM

UNION

The final stage of the mechanism forces a balance for the growth phase. It is akin to sleep and the unconscious, because it is a process that happens between the phases of growth and change. It's an ordering phase that extends from our commonly held notions of equilibrium. I believe that equilibrium can be explained in three slightly different, and somewhat conflicting manners: The first is by explaining its usage in physics as a function of disorder. According to the perspective of physicists, the Universe is always moving towards equilibrium and randomness; meaning that energy wants to be evenly dispersed across the universe. For this reason, two nearby gases of different temperatures are forced to equilibrate their particles. There is also a second notion of equilibrium that applies to intelligent structures. In this case, I believe that equilibrium represents some midpoint within a struggle between order and disorder. This means that as intelligent beings, we are constantly fighting the notion of complete disorder, and random dispersion. We are naturally forced by attraction into establishing an equilibrium between disorder and order. In the third case, as found in the fundamental mechanism, equilibrium acts to balance the distribution of physical particles with the distribution of patterns in mindspace, because the matter found within our environments must be reflective of pattern distributions. In all its cases, one fact about equilibrium remains constant: it describes the periodic management of an environment.

EXCHANGE

Transfer of patterns

NETWORK
between
organizations

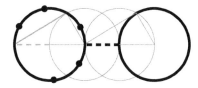

CHANGE
in the c-ratio of a
single organization

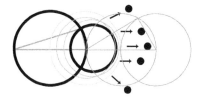

EJECTS
arrangements of
particles of matter

TRANSFER
of material to a
connected
organization

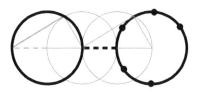

UNION NETWORK

As a symbol interacts with another symbol, its dyad fluctuates, and so does its c-ratio. In turn, the fluctuations in complexity force patterns to move between the symbols. Thermodynamics explains the displacement of particles between bodies on the small scale. However, large symbols in mindspace will exchange large masses of patterns that look like consumer products, documents, people, and other massive arrangements of particles. The goal of each symbol is to capture patterns to fuel its lifespan vis-a-vis any competitors. Note that organizations found to be stuck within any particular phase of growth or change, will appear uneconomical, frustrated, and will eventually waste its time.

ECONOMICS

Basic economic model of mindspace

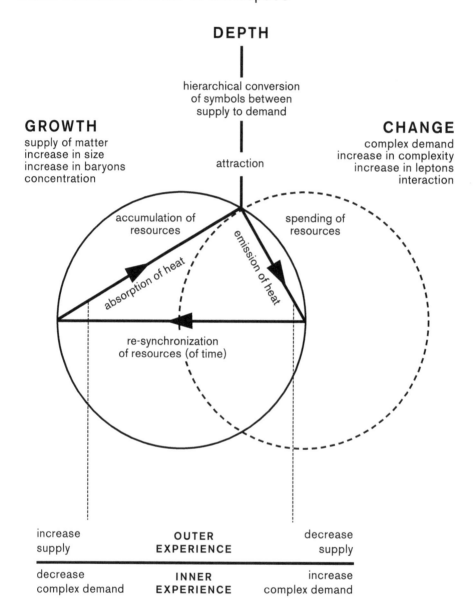

DEPTH

hierarchical conversion
of symbols between
supply to demand

GROWTH

supply of matter
increase in size
increase in baryons
concentration

attraction

CHANGE

complex demand
increase in complexity
increase in leptons
interaction

accumulation of
resources

spending of
resources

emission of heat

absorption of heat

re-synchronization
of resources (of time)

increase supply	**OUTER EXPERIENCE**	decrease supply
decrease complex demand	**INNER EXPERIENCE**	increase complex demand

Mental economics:
Relating to the production, development, and
management of material *and mental* wealth

UNION

One should imagine each fundamental mechanism in constant flux: its angles and ratios of connectivity are always changing. It contorts like a beating heart while emitting thought. The fundamental model of economy provides the basic perspective for understanding the interaction within a society of minds. It has a deep connection to thermodynamics and it preserves the laws of energy conservation across mindspace. It describes the economic process for all organizations equally, down to the atom, by explaining that during growth phases, an organization will absorb heat, build its time, and grow its matter. During a change phase, an organization will release heat (emotion), spend its time, and move matter around itself. The cycle of work is very much akin to a roller coaster. It could be said that all organizations follow the paths of interconnected roller coasters, and that our days are filled with highs and lows as a result of our intellectual motion across this mindspace geometry.

The fundamental mechanism can be used to explain all types of interactions, including particle interactions. But I like to use it to explain macroeconomic variability. Here are two of its consequences:

1 - If an unbalanced number of minds are demanding complex, kinetic energy, known as change, chaotic economic environments will rapidly surface.

2 - If an unbalanced number of minds are supplying matter, known as growth, noticeably slower economic environments will surface. Too much work without any change slows the motion of matter across our environments, making progress look slower.

Since people leverage this cycle while thinking, their dualistic leanings for political progressivism (change) or conservatism (growth) are powerful indicators of the demand-to-supply ratios (c-ratios) that exist across their section of mindspace. If a society is entirely progressive or conservative, this indicates a major distortion in equilibrium. These distortions result in cascading sets of problems that will eventually affect all interwoven arrangements of symbolic matter. A society's problems are then blamed on 'this or that', when in reality all problems are caused by unwanted cascading events resulting from the chronic disruption (frustration) of the fundamental mechanisms that balance the hidden realm shared between our minds.

One day man will connect his apparatus to the very wheel-work of the universe [...] and the very forces that motivate the planets in their orbits and cause them to rotate will rotate his own machinery

— Nikola Tesla

MECHANICS
CHAPTER VII

ENGINE OF TIME

ORDER
rotate

INTELLIGENCE
pull and push

The supremacy of non-physical reality was first suggested by the ancient philosophers, Plato and Socrates. They were adamant that the Universe was dualistic and extended into an abstract realm. They reasoned that thought was fundamentally different than matter, and thus a separate region of the Universe was necessary to manage it. They went so far as to reason that all matter must be somehow bound to this hidden region, and that this region was the essential origin of all natural things. Their theories on the subject were vague and often made with respect to a 'soul' and thus thousands of years later, their ideas mostly sit idle, ignored by the majority of the scientific community. However, I have shown that their basic concepts can be re-examined with fresh eyes, expanded with modern vision, and reframed into a comprehensive worldview.

This chapter will introduce a mathematical architecture used to contain symbols within pockets of time in space. I believe it explains how a mathematical world of forms can bind our bodies with our minds. The physical space and the abstract space are built using similar principles of structure, however there are some important differences between the two. One of those differences is that symbols in the mindspace architecture are all meaningfully interconnected through degrees of separation, while systems in the physical space appear disconnected. Mindspace doesn't have any empty space, while the physical universe does. To plot the physical universe one would use vectors across a Euclidean grid of rational numbers, but to plot mindspace one would use spinors across a complex plane. Despite their differences, the Pythagorean theorem unites both spaces and the IO diagram models the complex energy mechanism that manifests growth and change in both spaces simultaneously.

It should come to little surprise that our minds are based upon sacred geometries. For thousands of years, we have looked out into the world and repeatedly described the same numerological and symbolical associations between simple shapes and meanings, because our minds are built upon geometric structures that provide our windows into reality. This chapter will go into depths explaining the connections between irrational constants, their basic concepts, and some specific geometries that arise from them. I believe that our minds are laid upon vast numbers of sacred geometries, and the geometries combine to form a machine that motivates symbolic pockets of time across the layered grids of mindspace which eventually forge the material world around us.

Consider that we once thought that π was 3. And then, through technologies like the compass and square, we found that π was actually closer to the square root of 10. Later, we discovered it to be 22/7. One day we found that π is not 3.142, but actually 3.1415. With modern computers we know millions of digits of π, but without tools, our conscious minds can't differentiate between 3.1 and 3.14. Its reason is fairly simple: all minds truncate information they sense as to optimally position it within a mental map. A mind gathers and fits information within its map, like a well-fitted puzzle, and for this reason, our senses are inherently limited. The mental maps that we each assemble become part of mindspace - they are mirror-symmetries of the physical world within a complex grid of spacetime. This chapter will illustrate the fundamental architecture of this truncated, geometrical mental grid. Each element in the grid is constructed by the 5 facets of energy (ie. relationship patterns) within a 12-dimensional space. They form geometries that build into invisible engines that bring the mind-matter mechanism to life. If you've ever wondered how the human brain really works - I think this is it how - and it's not like any computer you've ever seen before. One day, we will exert a great deal of control over the mechanized hyperspace that exists within and between our minds and brains.

SUN SYMBOLS

Re-imagining the 5 patterns

HIERARCHICAL PATTERN
(Fundamental diagram)

UNION PATTERN
(Venn diagram)

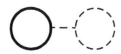

NETWORK PATTERN
(Network diagram)

SYSTEM PATTERN
(OI diagram)

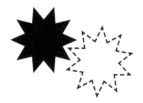

MECHANICAL PATTERN
(Gear diagram)

As I have shown throughout this book, this entire theory is based on these five patterns, and they pre-date all life in the Universe. As a result, each type of pattern can be conceived as metaphor for the stars, adding an unexpected twist to Carl Sagan's famous phrase: 'We are made of star stuff.'

The Sun is 99.8% of the mass of our entire solar system. All civilizations, and the religions that bound them, have been founded on Sun worship because the Sun is the most attractive symbol, sharing its energy with all people.

The Sun is physically responsible for the formation of all life on Earth by emitting light and heat. The deeper truth is that humans owe everything to the Sun, as do all solar minds. The Sun dwarfs everything, including the planets, by such measure as to almost signify ownership over them, yet its love is unconditional, unstoppable, and everlasting; these are its values that are taught in religion through references about the 'Son' of God. The planets that orbit the Sun are its followers and they cycle it in a way that is symmetrical to the way symbols cycle the human minds in mindspace. It is not to say that thought follows the structure of the solar system, or that the solar system follows the structure of thought - it's to conclude that both perspectives are part of a symmetry that transcends time and scale.

It is particularly striking that the five relationship patterns can be visually related to the five most important designs of the Sun. It is interesting because the relationship patterns are the sources of energy in mindspace, and stars are our most important sources of energy in our physical space. Now the question arises, how is this coincidence possible? Thought theory answers it in this way: human logic is meant to decipher forms of energy and the Sun has always been our ultimate depiction of energy. Since the patterns pre-date the Sun (and all other stars), when we draw the Sun, our unconscious minds are instead drawing out the fundamental patterns of mindspace. Although we consciously think we are drawing the Sun, I believe our unconscious is yet again manipulating our behavior in an effort to create a balance between abstraction and physical reality. Take note that our minds are therefore built upon a framework of stars.

As such, the Sun is a symbol of thought and thinking, and it's very likely that all highly intelligent beings (ie. aliens) would come to the same conclusion concerning their own minds and their own Sun.

Symbolic features

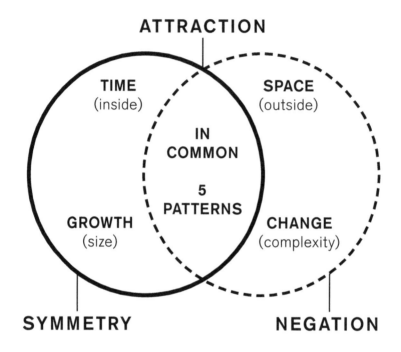

UNION

In the previous chapter I used the union pattern to explain the origin for the most basic principles of physics, but it can also be used to explain the origin for the basic principles of the mind. The Universe uses conceptual layering to create interrelated diversity, and all living beings inherit knowledge directly from the dyads that structure their minds.

EVOLUTION
Hierarchical patterns

DARWINIAN EVOLUTION
'SURVIVAL OF THE FITTEST'

Applies to life
      energy
      neurons
      thought
      corporations
      societies
      everything

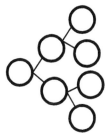

According to Dan Dennett, there are exactly 3 laws guiding evolution.[1]

Replication
Variation (Mutation)
Competition

PRINCIPLES OF EVOLUTION

Symmetry
Attraction
Negation

PRINCIPLES OF MINDSPACE

HIERARCHY

The mindspace operates by the same principles governing Darwin's theory on natural selection: 'Only the fittest survive.' Since fitness is a function of relationships held, all symbols are subject to natural selection, and as a result, all material arrangements are also subject to it. A relationship is the most basic form of power for all living and non-living things, because it is capable of affecting the speed at which time flows. The purpose of the mindspace grid is to maintain relationships (to control time).

HIERARCHIES

Organizational division

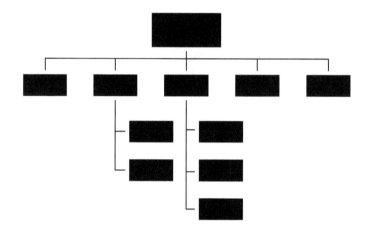

HIERARCHY

Organizations utilize their complex substructures of relationships to interact and move across mindspace. As they interact, they merge, build, and appear to destroy each other. I think it can be said that life is birthed when a mind develops self-will, and organizations that act upon the distinction between themselves and their environment demonstrate willfulness. How does the architecture of mindspace generate will, and does will even exist? Do the rules guiding the unconscious mind create will for each of us (as the unconscious mind displaces symbols according to some metaphysical laws)? Also, should will power be increasingly considered an environmental and societal issue, instead of a strictly personal one?

THEORY OF THOUGHT

LIVING FORM

Specially designed

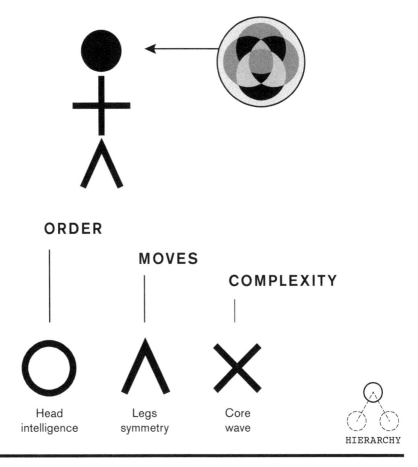

ORDER

MOVES

COMPLEXITY

O	∧	X
Head	Legs	Core
intelligence	symmetry	wave

HIERARCHY

Could the physical body have evolved thousands of different ways? I think that the human form is a reflection of the fundamental diagram and its three basic shapes of mirror-symmetry. The entire shape of man is patterned on the dimensions of mindspace, and it also seems that all animals on Earth are similarly bound. As such, each living body is a complex organization that evolved perfectly with respect to mindspace, which makes it clear to me that there must be a hidden fabric that binds all living creatures together, but how is this fabric mathematically described?

We tend to assume that space has infinite possibility and the irrational constants help us to believe in infinity, because these special numbers have never ending decimals. But what is infinity and can a number that we use every-day be infinite? The truth is that irrational numbers make little sense. How can the border of a circle never end? How can the diagonal of a square be infinitely long? For these reasons they are called irrational. They defy logic. Therefore all mathematicians must round or truncate irrational numbers to use them in ratio-nal calculation. So in truth all numbers must be shortened before they can be used in calculation. The only difference between what I'm going to show you and what mathematicians do is that I choose to truncate irrational numbers at earlier digits than a mathematician normally would.

Now, I'm going to make the case for early truncation. Let's say that I have two numbers, 15.25 and 15.35. These numbers are not exact. But if I truncate the .25 and the .35 from each number respectively, I'll have two number 15s, which are now exact. People do this all the time in their minds. For example, people will often remember the price of $15.25 as $15. It's easier to integrate truncated numbers into one's brain. It's also the reason why prices usually end in .99.

So what's the connection between a mind and a universe? Let us re-think both in this way: imagine each was a surface of pixels, like a computer screen. To reach the 15th pixel, a signal must be sent 15 pixels over. But let's say a signal is sent 15.8 pixels, what happens? Well, the 15th pixel would light up, and the 16th would not, because 15.8 doesn't quite reach the 16th pixel. Essentially, the signal length would be truncated to 15 and another number such as 15.9 would also be truncated and only light up the 15th pixel. The numbers 15.8 and 15.9 could be considered equal in this pixel based world. To light up different pixels for 15.8 and 15.9, by differentiating their values, the screen would need a higher depth of pixels. There would have to be more pixels within pixels (ie. fractal). Therefore, the 'depth' of space comes from each additional decimal and is conceptualized by the increasing density of pixels within additional layers.

What I've discovered is that the mind (and Universe) was founded across four layers of inner-pixels. It did so because the infinite depths of pixels did not yet exist when it came into creation (before inflation). This forced any irrational numbers to extend across one to four scales of depth, and then stop. Through this scheme, irrational numbers were assembled into elegant geometries that I believe can be identified at the very foundation of existence. I also believe that these geometries explain the basic framework of the mechanism for assem-bling the mindspace grids which integrate into a universal consciousness.

*Nothing exists
until it is measured*

— Niels Bohr

TRUNCATION

An irrational form of mathematics

NOT TRUNCATED

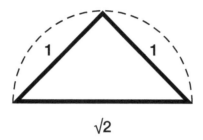

TRUNCATED

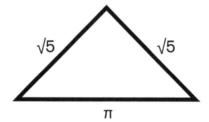

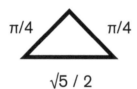

The 45°-90°-45° triangle above was highly revered by the Pythagoreans. They were among the earliest in Greece to realize its exact relationship with the square root of two, which was found to be irrational and quickly dismissed as magic by their contemporaries. When truncated, I have discovered that the √2 has a special relationship with π, Φ, and √5. In thought theory, varying scales of this basic right triangle can be used to relate important irrational constants together. Most mathematicians would judge my calculations to be inaccurate, however I intend to prove that there are significant connections to be made between irrational constants at truncated scales.

LUNE OF HIPPOCRATES

Converting between the physical and the abstract

PHYSICAL
Equal in numerical area (.5)

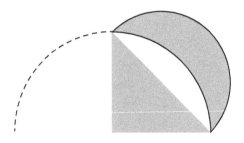

ABSTRACT
Equal in abstract area (half)

 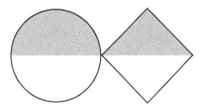

The Lune of Hippocrates was recorded around 450 BC. It was found while comparing two fundamentally different objects: the circle and the square. How does one compare a circle to a square in words? It's essentially like comparing apples to oranges. But regardless that's what our minds do. We find ways to compare and value vastly different symbols. If someone says that Jim is half the man that his friend Joe is, how can such a comparison be based? Where do the symmetries (comparisons) take place? This diagram may provide an explanation for converting concepts into geometrical symmetries within the mind.

SYMBOLIC SHAPES

Circle and square coincidences

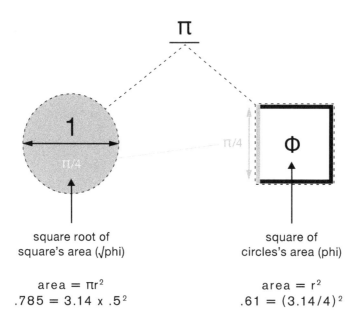

$$\frac{\pi}{}$$

$$1$$

$$\pi/4$$

$$\pi/4$$

$$\Phi$$

square root of
square's area (√phi)

area = πr²
.785 = 3.14 x .5²

square of
circles's area (phi)

area = r²
.61 = (3.14/4)²

People have been trying to square the circle using a compass and straight edge for a very long time. Squaring the circle implies creating a circle and square with the same area. Pi's irrational and transcendental value is the obstacle in doing so. However if we use truncated mathematics, we can discover a very special coincidence in geometry between the areas of a circle and square that each have a perimeter of π: the shapes can be related by φ. In effect, a circle of any size always contains 78.5% of the area within any square that inscribes it. It turns out that the square of 78.5% is 61%, implying that the circle and the square can be related by the golden ratio within a truncated space. Considering this convergence along with this diagram's coincidence in lengths, areas, and the number 1, it strikes me that there must be a hidden, sacred geometry embedded at the truncated scales of mindspace.

APPROXIMATIONS

Equations using truncated irrational constants

$\gamma = 0.577$, $\varphi = 0.618$, $\varphi^2 = 0.785$, $\sqrt{3}/2 = 0.866$, $\sqrt{\Phi} = 1.272$

$e/2 = 1.359$, $\sqrt{2} = 1.414$, $\Phi = 1.618$, $\sqrt{3} = 1.732$, $\pi/2 = 1.570$

$\sqrt{5} = 2.236$, $e = 2.718$, $\pi = 3.141$

$(1/2)^2 \approx \varphi^2 - \varphi^4$

$1/2 \approx \sqrt{2} \times \varphi^2$

$\gamma \approx (\pi/2) \div e$

$\varphi \approx (\pi/4)^2$

$\varphi \approx (e/2) \div (\sqrt{5})$

$1 \approx (\Phi/\sqrt{2}) \times (\sqrt{3}/2)$

$\sqrt{\Phi} \approx 2 \div (\pi/2)$

$e/2 \approx (\sqrt{3}/2) \times (\pi/2)$

$\pi/2 \approx \sqrt{\varphi} \times 2$

$\pi/2 \approx 1 + \gamma$

$\Phi \approx \sqrt{2} \times (\pi/e)$

$e \approx \sqrt{3} \times (\pi/2)$

$\sqrt{5} \approx \sqrt{2} + (\sqrt{3}/2)$

$\sqrt{5} \approx \sqrt{2} \times (\pi/2)$

$\pi \approx \sqrt{2} + \sqrt{3}$

$\pi \approx \sqrt{5} + (\varphi \times \pi/2)^2$

$\pi^2 \approx (\pi/2)^2 + e^2$

$\pi^2 \approx \Phi^2 + e^2$

I've collected this list of approximations over the last several years. This collection is based on rules: 1) Every equation uses at least one irrational constant; 2) In calculation, every constant must be truncated between its 2nd and 4th digit, inclusively. A constant is not allowed to be truncated at its first scale because the hierarchical pattern must include more than one scale; 3) Equations can be grouped and illustrated as geometries. Something happens when applying variable truncation - the equations produce answers within fuzzy ranges. For instance, this might mean that a result may be somewhere between 0.6 and 0.625, given the choices of scales. However, there will be at least one choice of scales that produces the best approximation, which may be for example, 0.618 (φ).

MIND-MATTER SPACE

The center of creation

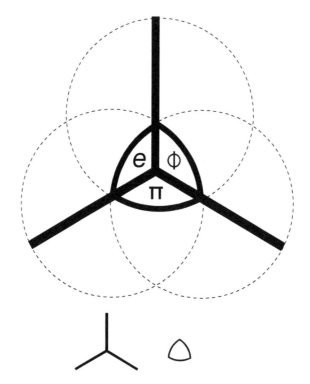

INFINITE SPACE FOR 3-DIMENSIONAL PHYSICS (REAL)	**TRUNCATED SPACE FOR GRID OF SYMBOLIC PATTERNS (COMPLEX)**

HIERARCHY

These hidden geometries appear to point to an abstract region that exists across every point in space. Symbolic structures emerge from them that are built upon the connections between circles, triangles and squares. I will try to show how these elegant shapes converge into an 'engine' that move patterns in mindspace before any bits of matter are forced to move in physical space.

ENGINE MECHANICS

A symbolic engine

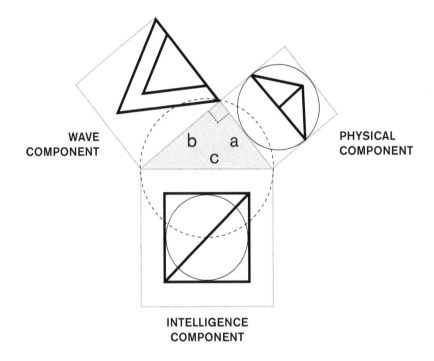

**WAVE
COMPONENT**

**PHYSICAL
COMPONENT**

b a

c

**INTELLIGENCE
COMPONENT**

$$c^2 = a^2 + b^2 \qquad 1^2 = .6^2 + .8^2 \qquad 1^2 = (3/5)^2 + (4/5)^2$$

φ   Φ/2

HIERARCHY

In this important diagram, the Pythagorean theorem joins 3 squares into a right triangle, and in thought theory, each square is said to contain a truncated shape that represents one of the three families of dimensions. The Pythagorean theorem is the base of a symbolic engine that draws a circle in hyperspace, and the circle that emerges is not just a circle, but a hyperspace container embedded with both physical and abstract dimensions.

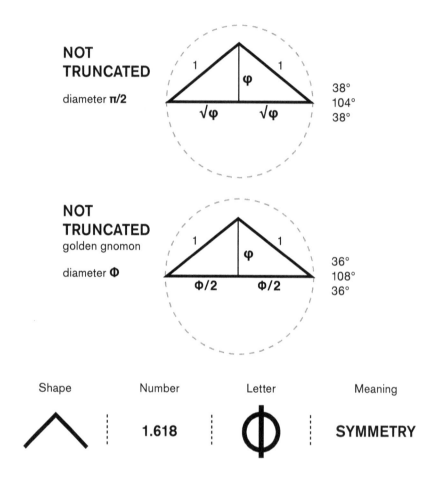

NOT TRUNCATED

diameter π/2

1   φ   1

√φ   √φ

38°
104°
38°

NOT TRUNCATED

golden gnomon

diameter Φ

1   φ   1

Φ/2   Φ/2

36°
108°
36°

Shape	Number	Letter	Meaning
⋀	1.618	Φ	SYMMETRY

Using truncation, these two shapes can be equated together and some valuable relationship between pi and phi emerges. This particular shape is the component of the symbolic engine that represents the **physical dimensions**: time, height, width, and depth. It also represents the **first law** of thermodynamics: *heat and work are forms of energy transfer*. It's synonymous with the change phase and it's responsible for transferring information across the mindspace grid.

INTELLIGENCE

Assembling a mind

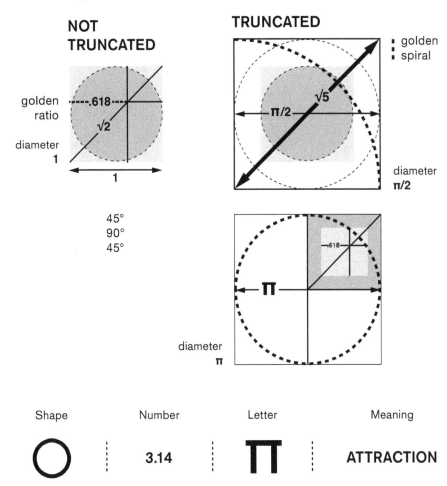

NOT TRUNCATED

golden ratio
.618
√2

diameter 1

1

45°
90°
45°

TRUNCATED

golden spiral

√5
π/2

diameter π/2

618
π

diameter π

Shape	Number	Letter	Meaning
◯	3.14	∏	ATTRACTION

This geometry represents the **dimensions of intelligence** and is respon-sible for creating individual positions on the grid. It relates to the **second law** of thermodynamics, explaining irreversibility. The law states that *the entropy of any isolated system not in thermal equilibrium almost always increases, and never decreases.* The second law of thermodynamics explains that the depth of information across a grid always increases and never decreases until some equilibrium is reached.

WAVES

Growing the mind

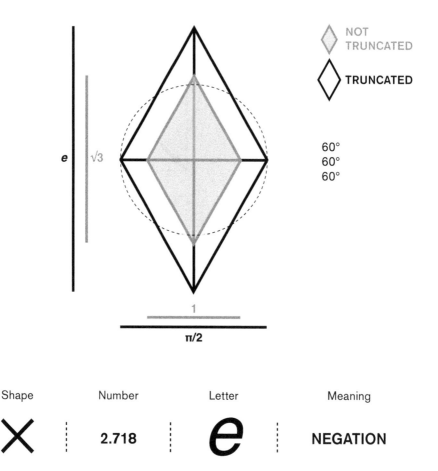

Shape	Number	Letter	Meaning
✕	2.718	*e*	NEGATION

The **third law** of thermodynamics states that *the entropy of a perfect crystal at absolute zero is exactly equal to zero.* Coincidentally, this component's shape looks like a crystal and it's responsible for managing the growth phase of each cycle. In thought theory, it is hypothesized that this component wiggles back and forth, forcing energy across different layers of grids. The wave energy goes on to power the rotation of the physical component across an individual layer. This particular component represents the **wave dimensions**: period, amplitude, frequency, and wavelength.

EQUILIBRIUM

Balancing mind and matter

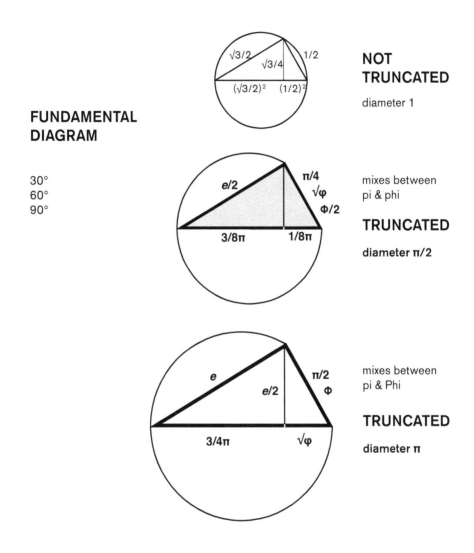

**FUNDAMENTAL
DIAGRAM**

30°
60°
90°

**NOT
TRUNCATED**

diameter 1

mixes between
pi & phi

TRUNCATED

diameter π/2

mixes between
pi & Phi

TRUNCATED

diameter π

The **zeroth law** of thermodynamics states that *if two systems are in thermal equilibrium with a third system, they must be in thermal equilibrium with each other.* The fundamental diagram is at the center of the symbolic engine, and it's a mechanical frame that joins three geometrical components in order to drive the growth, change, and final arrangements between organizations.

MECHANICAL PARTS

Sacred blueprints

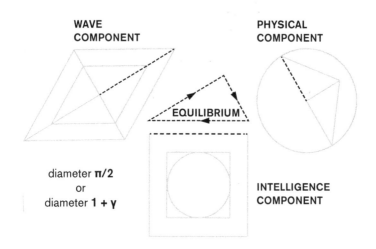

WAVE COMPONENT

PHYSICAL COMPONENT

EQUILIBRIUM

diameter **π/2**
or
diameter **1 + γ**

INTELLIGENCE COMPONENT

MECHANICAL ENGINE

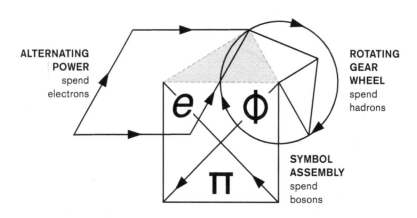

ALTERNATING POWER
spend electrons

ROTATING GEAR WHEEL
spend hadrons

SYMBOL ASSEMBLY
spend bosons

e ϕ π

The three components interlock into the mechanism (engine) using one of their shared lengths, and each one represents a family of dimensions that have a 'diameter' of 1.57. This value should relate to π/2, and it should imply some connection to half periodicity (of time).

EULER'S CONSTANT

Harmonic series $\Sigma(1/x)$ - natural logarithm $\ln(n)$

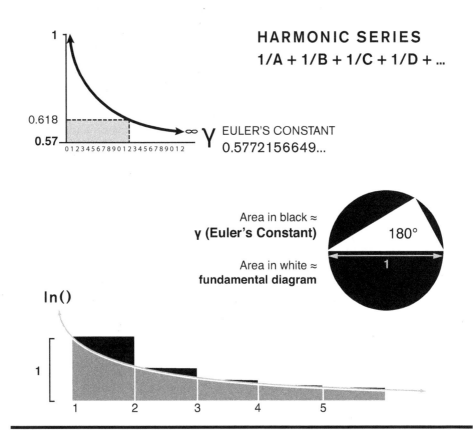

HARMONIC SERIES
$1/A + 1/B + 1/C + 1/D + \dots$

γ EULER'S CONSTANT
0.5772156649...

Area in black $\approx$
γ (Euler's Constant)

Area in white $\approx$
fundamental diagram

180°

$\ln()$

Euler's constant (γ) is another remarkable irrational constant. Its premise is fairly simple: like Φ, it results from a mathematical series. The harmonic series is an infinite progression of increasingly smaller rectangles. To calculate Euler's constant, a parabolic curve, known as the natural logarithm, is overlaid onto the series. The areas of the rectangles above the curve are added together, and this sum slowly converges towards Euler's constant, 0.577... The question arises, what is the meaning of the area above the natural logarithm? It turns out that Euler's constant crosses 0.58 into 0.57 (from the equation in the title) at exactly the 180th value in the series. 180° is equal to both, a half of a circle ($\pi/2$) and a triangle. This is important because at the 180th value, Euler's constant can be truncated (.57) to coincide with ($\pi/2$)-1. It can also be shown that the left over area of a circle, after inscribing within it the shape of a fundamental diagram, is approximately equal to Euler's constant. It appears that the metaphysical purpose of this critical constant is to assemble a system into a state of equilibrium, inscribed between a circle and its fundamental diagram.

I am illustrating concepts that are very difficult to explain and that would be practically impossible with words alone. In brief, I believe my research points to a hidden region of space that we've been overlooking for millennia. The reason we've ignored it is that we don't need to know about it to produce more accurate results in physical space. We don't want to use truncated numbers because they are less accurate. They don't help us whatsoever when measuring the distance to Mars for instance. But what I do believe is that these numbers are very important for designing an architecture that mimics the brain, because a brain isn't an infinite region like the physical space of the cosmos. It has a finite number of parts that must approximate the Universe, and its working parts must truncate information on some level. I believe that a brain will truncate all organizations of information that are inputted into it and it will do so at a very short depth (after a few scales of organizational hierarchy). It doesn't mean that a brain can't remember 3.14159265359, but it will only do so if the size and complexity of its mental map treats that number as a short enough distance (in terms of symmetry).

To explain it a different way - consider a bios. A bios is a component that is absolutely necessary for making a computer function. The bios is a chip (integrated circuit) that stores a basic set of instructions assembled by 1s and 0s, within a very limited memory bank. In fact, the built-in memory bank of a bios only has to be as large as the instructions need it to be. It would be uneconomical to build a memory bank that exceeds the storage needs of the instructions. If the instructions call for the division of two numbers, and the result is a number with an infinite number of decimals, the memory bank must truncate digits to fit some value into storage. What this means, is that a bios will not have enough room to store huge numbers, so it naturally *truncates* them, because *rounding* might store values that are numerically incorrect.

The basic building block for the brain, and the universe, is very similar to a bios. Instead of using 0s and 1s, the mindspace namely uses the monad and dyad, which in turn yield irrational constants, such a π, φ, $\sqrt{2}$, $\sqrt{3}$, and $\sqrt{5}$. Therefore, the universal bios that creates mindspace primarily calculates irrational numbers (and some important integers), and like a computer's bios, it cannot store their infinite values. So the mind truncates them at a surprisingly short depth (ie.

depth of the Tetractys). *Note that the baryonic strangeness of matter may arise from a basic process of truncation.*

It should be quite astonishing that by truncating irrational constants, very elegant geometries arise interconnecting them. These possible geometries were presented to you, over the preceding pages, and each one has striking coincidences with several of the most important irrational constants, integers, shapes, and principles. Did you notice how each component has a width of π/2? Did you notice how each shape relates to a revered triangle? Did you notice how each component can be connected to the fundamental diagram? Did you notice how each one corresponds to a basic law of thermodynamics (which explains the amount of work produced by machines and engines)? What are the odds of all of these converging coincidences?

Without complex measurement tools that peer across more than 2 to 4 scales of depth, these geometries cannot be falsified - no one can say the calculations are wrong. These geometries are the foundation of abstraction itself, and might quickly disappear as space is inflated (e.g. during cosmological inflation). These geometries should have existed at the birth of the universe, and should still appear during the birth of every new symbol ever since. For a moment, the geometries do exist, and I propose that they connect, work, and produce some form of interaction across a completely abstract part of the universal fabric, forging balance between all events transpiring in nature.

In physics it is often said that motion acts upon time, however in thought theory it is more accurate to say that time acts upon motion. So instead of thinking that an engine will increase the speed of an object, instead consider that it increases the rate at which time flows around the object. The object doesn't age faster, but its environment does. This way of thinking opens up the possibility that we can manipulate the speed at which the future materializes within our environments, by fueling some engines. I believe that our minds work to manipulate the execution of time and cause future events to come about faster, because mindspace is a framework of sacred, geometrical engines that are fueled by relationship patterns.

SYMBOLIC CONSTRUCTS

Models of interaction

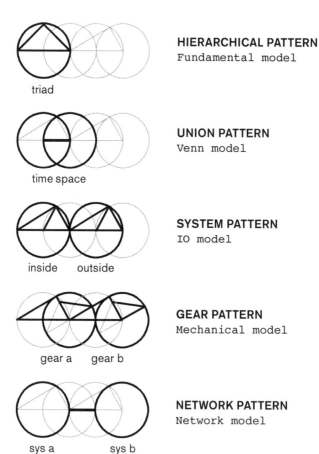

HIERARCHICAL PATTERN
Fundamental model

triad

UNION PATTERN
Venn model

time space

SYSTEM PATTERN
IO model

inside   outside

GEAR PATTERN
Mechanical model

gear a   gear b

NETWORK PATTERN
Network model

sys a   sys b

The five relationship patterns can be visually identified as a sequence during the interaction between symbols. Each symbol is an engine that combusts relationships, and all five patterns are needed to explain the full extent of the interaction between symbolic engines. The gear pattern will be more thoroughly described in my next installment of Theory of Thought - as there is much to be explained concerning it.

*When you concentrate
your energy purposely
on the future possibility that
you aspire to realize, your
energy is passed on to it and
makes it attracted to you
with a force stronger than the
one you directed towards it*

— Stephen Richards

COMBUSTION

Mass, light and motion

$$E = (m) (c) (^2)$$

**THREE COMPONENTS
OF ENERGY**

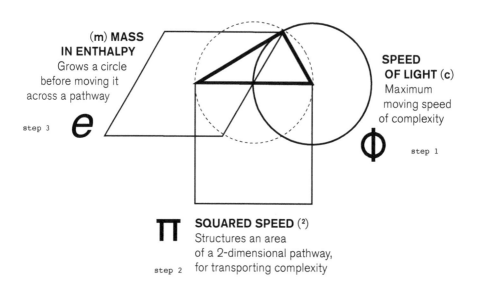

**(m) MASS
IN ENTHALPY**
Grows a circle
before moving it
across a pathway

step 3

**SPEED
OF LIGHT (c)**
Maximum
moving speed
of complexity

step 1

SQUARED SPEED (^{2})
Structures an area
of a 2-dimensional pathway,
step 2 for transporting complexity

Every symbol in mindspace is an **engine** representing an **open system** that moves time across a hyperspace. An engine can be of any size and as such, can represent one pattern or an arrangement of a trillion patterns. The maximum energy within an engine is calculated by $E=mc^2$, and each part on the right-hand side of this basic equation relates to a discrete component of a mindspace engine. An engine's maximum energy is therefore a function of its fastest operational speed possible (c), the square of this speed (^{2}), and the mass of patterns being propelled (m). In short, the parallelogram is a mass of time, the gear wheel accelerates it, and the square creates an area for it to travel into. Note: the diagram above is not showing the gear, only its wheel.

THEORY OF THOUGHT

VIS VIVA

The living force

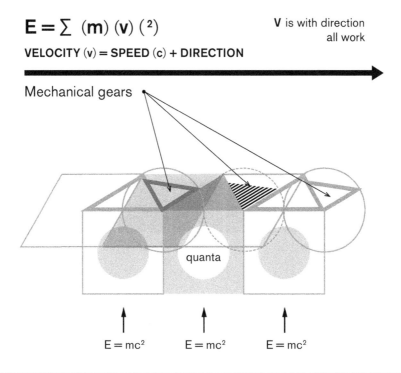

$$E = \sum (m)(v)(^2)$$

V is with direction
all work

VELOCITY (v) = SPEED (c) + DIRECTION

Mechanical gears

quanta

$E = mc^2 \qquad E = mc^2 \qquad E = mc^2$

The 'living force' refers to a perfectly elastic and frictionless mechanical flow of energy between systems in spacetime. In thought theory it refers to the flow of symbolic units of time between the geometric engines in mindspace. *Vis Viva* was eventually changed to include friction because all transfers of energy operate under some friction, which releases heat and wastes work. Therefore the total work produced from mechanical transfers is always less than the total amount of energy put into them. To describe *Vis Viva* in mindspace, engines are stacked horizontally into a grid and the distances between them are measured as rotations across a complex plane. As an engine wheel spins, the gear will collide with connected engines across the grid. If the gears between engines are symmetrical, there will be less friction, and less time will be wasted. This is also hypothesized to be the simple process explaining the chain reactions of thought in the brain.

LINES OF INTERACTION

A grid of engines working together

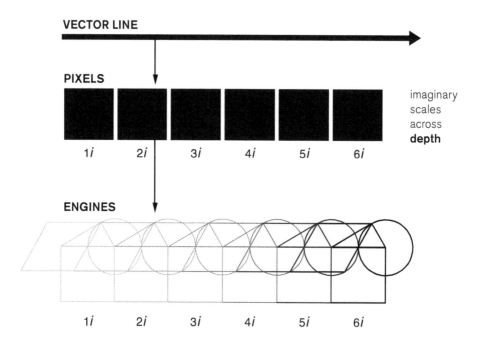

INTERACTION   Similar Scale (imaginary)
               Work

VECTOR LINE

PIXELS

imaginary
scales
across
depth

1i 2i 3i 4i 5i 6i

ENGINES

1i 2i 3i 4i 5i 6i

These interconnected squares simulate distances in mindspace. When the mechanical engines connect *horizontally* in a line of the same logarithmic scale (same size squares), they produce interaction. Changes in one organization can affect the physical growth and change of another organization, directly or indirectly, through these connected spaces. When time travels horizontally, work is performed efficiently and less time is wasted.

FRICTION

Heat is released to bridge new grids

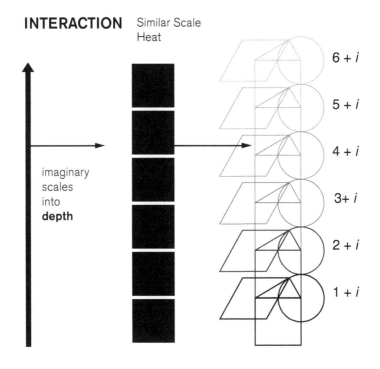

INTERACTION Similar Scale
Heat

imaginary
scales
into
depth

6 + *i*

5 + *i*

4 + *i*

3+ *i*

2 + *i*

1 + *i*

During interaction, the mechanical engines transfer a flow of time horizontally. However, the interaction is subject to friction, which wastes time. Friction's effect is to alert other engines in mindspace that some relationships are not optimally efficient from some lack of symmetry. So when friction takes place, some units of time (ie. photons) are released *vertically* in mindspace. Their purpose is to explore new pathways and lay the groundwork to bridge differ-ent symbols. This vertical interaction process does not result in direct work (i.e. efficient transfer of time), but it does encourage new interactions.

SCALES OF GROWTH

Increasing intensity

CONCENTRATION Different Scales
Logarithmic

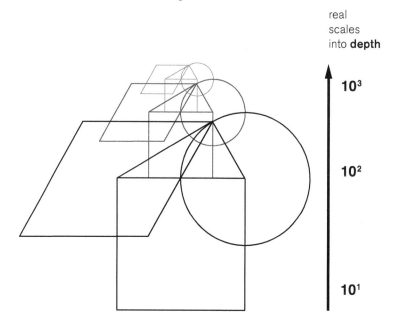

real
scales
into **depth**

10^3

10^2

10^1

Concentration is exerted by engines stacked across different logarithmic scales. These types of vertical stacks do not produce symbolic events (work), instead they provide *opportunity to grow events* by creating depth within an environment. For example let's suppose there are one million people in a network with one million connections between them. Now let's imagine that the network grows in density by a factor of 10 connections per person.

INVERSE SQUARE LAW

Increasing scale

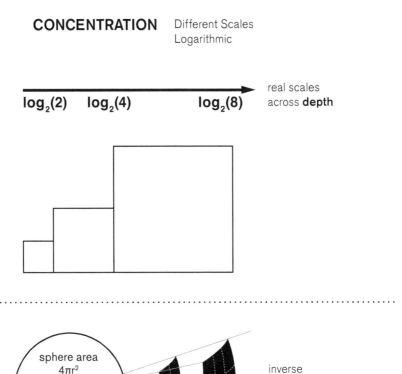

CONCENTRATION Different Scales
Logarithmic

$\log_2(2)$ $\log_2(4)$ $\log_2(8)$ real scales
across **depth**

sphere area
$4\pi r^2$

inverse
square law

The squared structure of the stacked geometric engines might explain why there is an **inverse-square law** in physics governing gravity, electrostatics, and radiation. This horizontal version of growth is inversely proportional to a growth in intensity, and it creates expansive organizations at the cost of intensity per scale. For example, instead of growing 10 connections per person, a network grows 10 people per connection.

RADIATION

State of acceleration/rotation

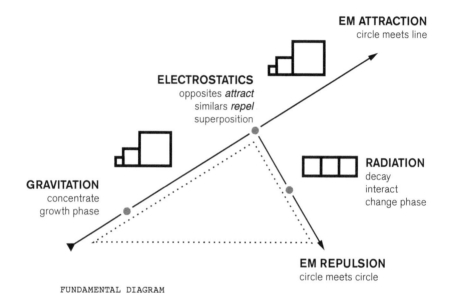

EM ATTRACTION
circle meets line

ELECTROSTATICS
opposites *attract*
similars *repel*
superposition

RADIATION
decay
interact
change phase

GRAVITATION
concentrate
growth phase

EM REPULSION
circle meets circle

FUNDAMENTAL DIAGRAM

As thought theory explains, mindspace is a web of circles traveling across lines to reach other circles. The distance of a line is equal to one rotation of the fundamental mechanism, and when the rotation ends, it could be said that a circle has reached its destination. Therefore, the growth phase starts the rotation of the circle, and the change phase generates its length. If the growth phase continues indefinitely, the circle grows into infinitely large logarithms of size. And without both phases, a cycle never ends and a perimeter is never measured. It must be also considered that every circle contains inner-circles, which travel their own lines (and fm cycles). So as a circle travels, it radiates information as its inner-organizations complete their own shorter cycles. There is even a larger perspective, where a circle represents an organization of matter, and as it grows in size, it accumulates more matter to eventually form into large physical environments. I believe this is the process by which atoms or any other form of complex arrangement is formed by cycles of the fundamental mechanism.

Ref. page(s): 192

TRUNCATION
Infinity within truncated geometries

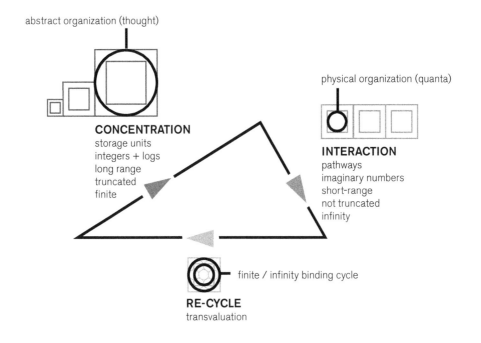

abstract organization (thought)

CONCENTRATION
storage units
integers + logs
long range
truncated
finite

physical organization (quanta)

INTERACTION
pathways
imaginary numbers
short-range
not truncated
infinity

finite / infinity binding cycle

RE-CYCLE
transvaluation

The cycle of the fundamental mechanism transpires across 1) logarithmic scales of depth, and 2) across an imaginary plane found at every individual scale. This means that the entire mechanism transpires across a complex space. During concentration, hierarchies of patterns are absorbed, and at some point in the phase, an interaction phase is triggered that uses those hierarchies to act upon other fm cycles located on the same imaginary plane.

Although this diagram might be used to explain physics, I think it should be even better suited at explaining how thought is organized in the brain. Between the concentration and interaction phase is an unconscious transvaluation phase. This phase of the fm cycle forms an environment, produces value, and manages aging. It truncates infinite lengths of time into discrete units of sacred (irrational) geometries, which position a mind on an infinite grid that sustains its abstract lifespan (ie. aging).

EULER'S IDENTITY

Creating time

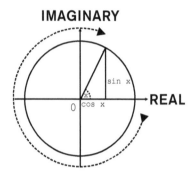

IMAGINARY

sin x

REAL

0 cos x

**fm cycle
across
mindspace**

intelligence dimensions
(sum/insert)

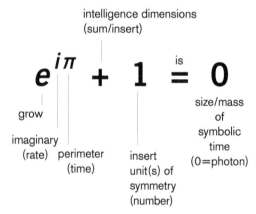

$$e^{i\pi} + 1 \underset{\text{is}}{=} 0$$

grow

imaginary
(rate) perimeter
(time)

insert
unit(s) of
symmetry
(number)

size/mass
of
symbolic
time
(0=photon)

**inner
workings
of the
cycle**

SYSTEM

Richard Feynman once said that "Euler's formula is our jewel... It's the most remarkable formula in mathematics". In thought theory, it is hypothesized that this derivative of Pythagorean's theorem is used to calculate an fm cycle's power across the
complex plane. Using this formula, each mind adds symmetries during its fm cycles
in order to manipulate the amount of inner-cycles that it generates. As more
symmetries are included within a single cycle, an engine increases its 'horsepower',
which leads to a greater influence over some environment (neighboring engines).

NUMBER LINE

Negative and positive space

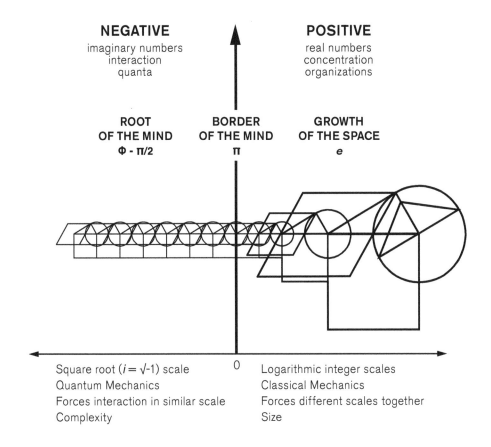

NEGATIVE
imaginary numbers
interaction
quanta

POSITIVE
real numbers
concentration
organizations

ROOT
OF THE MIND
Φ - π/2

BORDER
OF THE MIND
π

GROWTH
OF THE SPACE
e

0

Square root ($i = \sqrt{-1}$) scale
Quantum Mechanics
Forces interaction in similar scale
Complexity

Logarithmic integer scales
Classical Mechanics
Forces different scales together
Size

Interaction and concentration, which are 2 separate phases, can be basically illustrated as a single line. Note that $\sqrt{-1}$ is an imaginary number, and it represents a rotation, in contrast to a real number that represents a translation. Each cycle of the engine is distributed across these two perspectives of motion atop a single complex plane.

MOVING MATTER

The purpose of mindspace

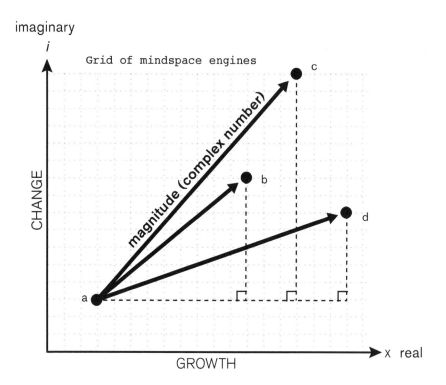

The number line from the last page can be even more simply illustrated as an x-*i* grid of complex space. In physics, each particle's wavefunction, or group of particles, is defined by a complex number. My conclusion: the Universe is a hyperdimensional grid **of dyads** used to displace symbols and wavefunctions, in a metaphysical and physical sense, respectively and simultaneously.

Let's say that I have my hand pointed at an empty space on a table. And I tell you that a particle will appear there. It will happen, I just don't know when. Alternatively, I could say that a particle will appear now, however I won't know exactly where. This basic observation is explained by the uncertainty principle, and quantum mechanics allows one to calculate some probability to say that a particle will appear here and now. Physicists discovered that by using fundamental vectors called probability amplitudes (aka. wavefunctions), they could predict when and where a particle will appear, yet when examining one particle at a given time, the results appear probabilistic, which means, not quite exact.

Thought theory isn't really about dissecting the mathematics of single particles. It's more of an attempt at predicting the position of whole arrangements of atoms, stuff like consumer products. So let's say that I point at a table and proclaim that a pen will appear now, it might appear there and now, or it might not. From common sense, I know that a pen might appear if I controlled someone or something into placing a pen there. I might even place a pen there myself at the right instant, just to prove my prediction true. The way to calculate whether I can command a pen to appear at some location is by calculating enough symmetries that I share with my environment. For example, a symmetry would have been created if I had previously asked a friend in the room to hold a pen. Through these types of connections, called symmetries, arrangements of matter can be predicted to appear.

In quantum mechanics, the probabilistic command over particles is achieved by measuring the magnitudes of complex numbers. While in thought theory the command over arrangements of particles is measured by some magnitude in relationships (symmetry). To unify the physics of physical and abstract spaces, the quantum world and the everyday-world should be measured using the same types of probability amplitudes. Perhaps each scenario is explained differently, but both scenarios should be calculated using the same mathematical concept. The symmetries across mindspace should be converted into probability amplitudes, and by applying quantum mechanical formulas, it could be predicted when and where any arrangement of matter will appear in any environment of physical space with the highest degree of certainty.

ABSTRACT MACHINE

Mechanized grid of mindspace

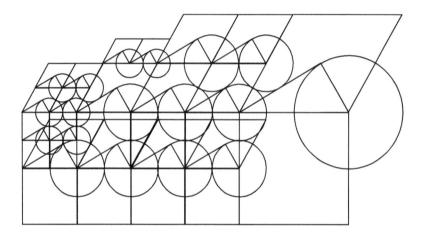

framework only partially shown; fractal in nature;
depends on final component & grid configurations;
resembles apparatus of pulleys and levers.

I believe that an entire mindspace is governed by symbolic engines that can be understood using modern laws of physics. Like a wavefunction, the machinery is entirely abstract, yet calculable. Symbols store and exchange energy between one another and the arrangement of physical space is manifested from them. When we're not paying attention, our unconscious minds use symbolic engines to order patterns across mindspace. But when we finally 'pay attention' our consciousness collapses the symbolic engines and the notion of materialism surfaces as objects are positioned to reflect the patterns they manifest from. As consciousness moves away from collapsed regions of the mind, the symbolic engines re-assemble themselves to once again manage the probabilities of symbolic events occurring in spacetime.

THEORY OF THOUGHT

WORLD OF SYMBOLS

Universal consciousness?

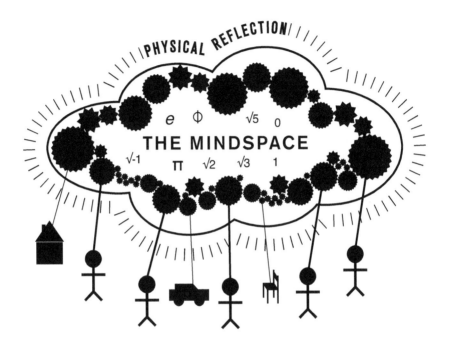

In the end, this chapter demonstrated that there should be a real, yet abstract space that exists binding mind and matter, designed with respect to the Sun. This world weaves matter, the brain, and the mind into a geometry of symbols that have defined structure within a hyperdimensional area. Symbols can be said to engage one another by exchanging circles of time across complex planes. According to the gear pattern, symbols can also be described as star-shaped organizations that extend deep into mindspace and the connectivity of their gears leads to a dynamic machinery of time.

It appears to me that some higher dimensions exist within the conventional grid-like frameworks often referenced in mathematics and physics. What I did, was to uncover the underlying components that empower the vectors of motion across the grids. These components create the order in our Universe and in our daily lives. They were speculated upon by ancient greek philosophers but until now we couldn't actually see them, because we had never identified them in this manner. This architecture should lend itself to uncovering the still hidden mysteries of science, including those assumed to remain within physics, neurology, psychology, and economics.

I believe that this book provides the basic framework for a new perspective on reality. It will help lead civilization to a more elevated position and it will help add clarity to the underlying purpose of complex systems and the unfolding events between them. I believe that this last chapter begins to explain the specific functions of the brain, and why neurons exist. It doesn't describe the brain exactly in a chemical and biological sense, but it does describe the natural foundation for what its biology is meant to achieve. Furthermore, it might be the only way of understanding new physics concepts like dark matter and dark energy. It might reveal the hidden variables that result in non-locality and entanglement. Ancient prophesies might actually be plausible given that a geometrical framework of symbols binds everything in the Universe. There could be periodic waves that cross these grids without our direct knowledge, yet affect a majority of our unconscious minds. This theory might also help us further conceptualize fractals, karma, chi, the paranormal, and even schizophrenia. What is being transferred in the mindspace, right now? What does it look like? Is there a build-up of destructive complexity leading to confusion and mental illness? How does a well configured mindspace lead to the perception of a well-functioning biology in a brain and body? Thought theory will eventually answer these questions using the inherent logic of relationship patterns. I believe that our philosophical views on thought and reality need critical restructuring, because our misaligned notions of them are the ultimate source for all our current, worldly problems. And I hope that soon, we will be better equipped at affecting change directly within mindspace, instead of focusing on the same material issues to little avail.

The energy of the mind
is the essence of life

— Aristotle

Reality is an illusion,
albeit a persistent one

— Albert Einstein

EPILOGUE

In the least, I think I've shown that there is a hidden aspect to the Universe that must be researched further. With an application of the rules of probability, I believe that this hidden, mathematical perspective will finally unify the intangible mind with tangible matter. For thousands of years, philosophers have hypothesized the union of mind and matter but have been unable to architect it using scientific doctrine. As such, we are still taught that the universe and the thought propagating our brains are separate entities. We are told that thought is produced from some forms of matter, when in reality this is not the truth. What we are told to believe is not an intentional deceit. There are some beliefs that are confused with truths and they are obstacles in our journey to becoming a great civilization. Marvelous events transpire when civilization crosses these obstacles. For example, some ancient intellectuals strongly believed that the Earth was spherical. However it could not be proven, and as such most people were taught that the Earth was flat, because it seemed to be flat. But then one day, Magellan circumnavigated the globe, and overcame an enormous obstacle. The New World was discovered in the process, causing an abrupt and powerful change in civilization. The obstacle that was overcome wasn't the voyage, or the boats, or the technology; the obstacle was a belief. Beliefs consistently hold back progress, and the only way of defeating them is by rigorous proof - equated to surrounding and conquering them from the outside-in.

Is this book similar to the notion of circumnavigating a globe? I'm not exactly sure what circumnavigating an abstract globe looks like, but this is the most interesting pathway around it I've found. Hopefully, this book is laying out a rough plan on how to surround and conquer a huge obstacle in perception. Maybe the plan is still incomplete. Some day, a completed plan will perfectly explain how to transform the flat, physical universe into a round, abstract mindspace. And maybe it will lead to a great expansion in many important beliefs while vanquishing the weak ones.

I will be the first to admit that this book surely contains flaws. I know of some already. There are aspects that I haven't examined closely enough, especially the concepts in the last chapter. I've been working on these ideas for over seven years and I consider some of them to be cutting edge. Without going further I cannot be aware that I may have inverted or wrongly connected some concepts. How do I know the engines work like they do? I don't. But I didn't guess at it either. All of my conclusions are approximations based on other tested approximations. I strongly believe that my foundation, based on the circle and line, and the three sets of dimensions that emerge from them has the tensile strength of a diamond, perhaps one still in the rough. From my perspective, I've joined together an abstract structure of such architectural integrity that it should hold up to the strongest of arguments and objections. I believe that my strategy in building from strongly rooted, conceptual pillars is why this theory is fundamentally correct. I truly feel that it points me into the right direction - one that I can deeply envision. There was an extensive amount of information that I had to integrate into a single framework, and the end result is polished, but still unclear, perhaps like the first magnifying lens. But unlike the first lens, my lens peers into an *abstract* universe. How can I deny what I see with it: abstract networks of organizations that exchange energy, pulling on minds and forcing bodies to move? How can I deny its perspective on materialism? How can I deny the existence of five supernatural patterns that basically govern the mind, brain, and material physics? Answering these questions and others like them, is what drives me and this work. I have additional diagrams tying in the structure of an atom, the fine structure constant, radioactive decay, General Relativity, the Poincaré lie group, and dozens more scientific concepts. How is it possible that all of these ideas can be seen coming together without fail, and where will this interwoven stream of ideas end?

I wrote this book to show myself that it is possible to change the world, and I plan to do just that; but I also need your help in doing it. One person can spark change in the world, but many people are necessary to successfully propagate it. My hope is that you can contribute by continuing to put the pieces of this puzzle together. It's not that difficult, but to succeed at it, you have to believe in what I've shown you. You have to see it all around you, and once you see it, you can contribute to making it reality. And once realized, civilization will finally overcome an enormous obstacle, a new world will emerge, and everything will change abruptly, once again.

FURTHER WORK

These are some proposed steps that our community can take to help expand this theory:

1) Construct computerized diagrams of the truncated space, and the sacred geometries embedded within. The geometries might have to be constructed within a simulated, limited memory bank, like one included inside a bios.

2) Assemble complete computerized databases of abstract relationships between all material arrangements and the environments they are found within. Wikipedia is a good start. So is Facebook.

3) Build a consistent theory on the brain using the five relationship patterns. Everything that goes into and out of our brains is translated into and out of these 5 patterns. The unconscious brain is always at work ordering them into a map and the conscious brain scans the map.

4) Fully integrate the standard model of physics into the notion of a mind-space. Ultimately this means that each wavefunction is rooted in a mindspace truncated geometry, and the interaction between geometries reflects all the subsequent rules of physics.

5) Determine exactly how to translate symmetries into probability amplitudes. The values of the amplitudes should be derived by measuring relationships within one or more relational databases.

6) Perfect the notion of the fundamental mechanism. It should be explored as the root of a universal cycle binding the mind with physics - and not through telepathy or some other mystical means - but through probabilities that bind the organization of a mind with the arrangement of matter within a specific environment.

7) Build a software/hardware based artificial intelligence based on the rules and structures of this theory. The glossary is a good start in understanding its potential functions, classes and variables.

8) Expand on the notion of abstract time and the material effects arising from the systemic gain or loss of it.

9) Definitively prove or disprove whether abstract mass and dark matter are derived from the same concept.

10) Incorporate a form of calculus that reframes wave-like functions in terms of non-uniform rotations. Also examine the potential of employing Quaternions and Clifford Algebra (G.A.) for constructing the abstract space.

11) Determine the direction in which the mechanism flows. It might actually flow counter-clockwise (left), and therefore many of my diagrams should be mirrored vertically (it's a trivial process). Regardless, I feel that explaining everything with respect to clockwise rotation was more intuitive for me.

12) Practice thinking more deeply about relationships and the impacts they have on our reality.

13) Follow Theory of Thought on the web to stay current with the scientific progress made to this task list.

MY SEQUEL

The research phase for this book took me several years. I had to accumulate and distill a great deal of information in order to create such a comprehensive perspective on science and philosophy. If you look closely at my complete body of work, including my poster, you'll see lots of figures and diagrams that I did not include in this book. If I had included them, this book would be close to double the size and contain even more complex information. Whittling work is a delicate and painstaking process for every artist - a balance between form and function must be achieved. Leonardo da Vinci famously stated that a work of art is never finished, only abandoned. With this first book on the topic of symbolism, I feel that I had to stop short, however, I also feel that I was able to present my logical and mathematical viewpoints without going too far. In many ways, my first attempt is a way at describing mathematics without using so many equations. I hope that mathematicians will appreciate the way in which my math is presented, through the use of a unified set of diagrams that in essence represents mathematics in the purest of forms, without the need of equations. Equations are for people who have to calculate, while diagrams of equations are for people that need to understand why the equations exist in the first place. Having said this, a future installment might be named Theory of Thought: Mathematics and provide the definitive proofs that mathematicians need to re-create mindspace using numbers alone.

My next installment will probably be about Time. Time is perhaps the greatest of concepts, and I believe that the manipulation of time will provide mankind with the greatest of benefits for controlling our presence in the Universe. There is much to be said about time, including its subjective and objective perspectives. It is known that time flows independently across every body and there is not a universal, single viewpoint on its flow - its flow is 'relative' as described in Relativity theory. I think there is ample space here to describe the intricacies of time including its flow across the human mind and its ability to force changes in the behavior of people. I've already developed a great deal of models, such as the IO model and the Gear model, that explain the creation and storage of time within abstract space. These models will be the focus of my following book. The gear model will show how time can be some- what 'folded' like origami, and employed like a mechanical gear that pushes up against external folds of time (like an enormous machine). It will also discuss the idea of an 'instant' of time and how minds of similar instants can work together, while minds with large differences in instants cannot. As I said previously, there is a lot more to be explained - and I want to present my ideas in a patient and digestible format.

Video is perhaps the best tools I have at my disposal for explaining a new world view. In this ebook, the quality of my videos are the weakest aspect of its design - although I think they deliver great value to the reader. In my next installment I need to work with videographers and animators, in order to really bring this theory to life on screen. Beautiful video is really expensive which is why I was not able to deliver better videos this go round. In order to afford awe-inspiring videos along with Theory of Thought's sequel, I will launch a crowdfunding campaign. Getting funding isn't easy in this field of work, and I hope that I can reach out to my fans for the support I will need to continue expanding this topic. I also hope to meet a few other like-minded people to work together on the cutting-edge of a new scientific body reminiscent of early 20th century physics.

Please follow us online, rate this work publicly, and discuss it with your peers to help keep Theory of Thought alive and on track.

GLOSSARY

INTRODUCTION

0.1 **Metaphysics:** An obscure science rooted within Plato's Theory of Forms. Its scientists search for evidence of abstract structures that exist independently of the brain.

0.2 **Physical Dimensions:**

> 0.2.1 **Time:** is the all encompassing dimension of physical space. Every direction in space travels across it.
> 0.2.2 **Height:** is the up and down direction.
> 0.2.3 **Width:** is the left and right direction.
> 0.2.4 **Depth:** is the back and forward direction.

0.3 **Spacetime:** is a mathematical model combining space and time into a single concept. It joins the first three physical dimensions - height, width, and depth - with the fourth physical dimension, time.

0.4 **Mindspace:** is a hyperspace architecture that joins the dimensions of spacetime with two other sets of 'higher-dimensions'. The first set of dimensions are used to explain the properties of waves: period, amplitude, wavelength, and frequency. The second set of dimensions are used to explain the properties of intelligence: meanings, shapes, numbers, and letters. Exactly three sets of dimensions - the physical, wave, and intelligence - are combined into a single concept, mindspace, that binds a physical space with an abstract space.

0.5 **Thought theory:** uses diagrams of mathematics and physics to explain the way in which symbols are structured by physical and abstract dimensions.

0.6 **Fundamental Mechanism:** is the most basic interaction within mindspace. The mechanism binds the three sets of dimensions into a continuous, indivisible cycle. Every organization of matter in the universe, down to an atom, is governed by the fundamental mechanism.

0.7 **Patterns:** are facets of the fundamental mechanism from which physical systems are materialized. Each type of pattern explains a type of relationship that binds the mechanisms in mindspace.

0.8 **Systems:** are any physical arrangements of matter within the context of spacetime. It can also be used in reference to an organization in mindspace.

0.9 **Symbols:** are abstract (and physical) arrangements within the context of mindspace.

1.0 **Physical Space:** is the perspective of mindspace (0.4) where matter resides.

1.1 **Abstract Space:** is the perspective of mindspace (0.4) where patterns resides.

CHAPTER 1

1.1 **Mind:** Minds are organizations in mindspace that have some arbitrary level of complex arrangement. This arbitrary level is valued by a given c-ratio within some scale. All arrangements of matter, on all scales, belong to at least one mind, however most arrangements belong to many minds. Minds are under the constant influence of attraction, and they instinctively connect patterns together. To do so, they evaluate symmetry, and force new patterns from existing relationships. A mind's decisions on how and where to connect patterns are always driven by attraction, and never done arbitrarily.

1.2 **Relationship Patterns:** Organizations in mindspace are part of one or more relationships and they exist and interact according to five 'relationship patterns':

1.2.1 **System:** a pattern that produces diagrams that visualize distant circles within circles.

1.2.2 **Network:** a pattern that produces diagrams that visualize distant circles bound by lines.

1.2.3 **Hierarchical:** a pattern that produces diagrams that visualize circles bound by lines, with respect to time.

1.2.4 **Union:** a pattern that produces diagrams that visualize unions between circles. E.g. Venn diagrams.

1.2.5 **Mechanical:** a pattern that produces diagrams that visualize spinning circles interacting with other circles using gear-like triangles.

1.3 **Circle:** is an organization of resources in mindspace. It represents any mindspace organization.

1.4 **Thought:** the abstract substance contained within any circle. It is a biproduct of the processes going on in mindspace and governed by the principles of order that influence and externalize the material world.

1.5 **Inner-Organization:** is an organization of mindspace that is considered to be inside of a larger circle. Visualized by a systems diagram.

1.6 **Outer-Organization:** is the most immediate container of any particular system. It is a larger mindspace organization that contains smaller mindspace organizations. Inner and outer organizations exist on different scales.

1.7 **Line:** is a pathway connecting two mindspace organizations for the purpose of exchanging or transferring sub-resources.

1.8 **Symmetry:** of a physical system is a mathematical feature that is preserved under some change in perspective. In thought theory, it represents the resources that are exchanged between mindspace organizations. It is a shared quantity.

1.9 **Complexity Ratio (c-ratio):** is the ratio of lines to circles within an arrangement of network patterns.

1.10 **Size:** represents the perimeter of a circle. It is a function of a sum or product between the collection of its inner-organizations (1.5).

1.11 **Complexity:** represents a line. The length of a line is a function of the differences between two circles, like within a network. Since complexity refers to the line connecting circles, it also refers to the assembly of relationships in mindspace.

1.12 **Hierarchy:** represents an arrangement of circles in which they are placed above (1.6) or below (1.5) one another.

1.13 **Dimensions of Intelligence:** are unique, and abstract qualities of thought. These dimensions follow a hierarchy (1.12).

> 1.13.1 **Meanings:** are used to categorize thoughts within a circle.
> 1.13.2 **Shapes:** are used to illustrate the shape of thought. (circle)
> 1.13.3 **Numbers:** are used to count circles.
> 1.13.4 **Letters:** are used to describe distances (lines) between circles.

1.14 **Tree Network Diagram:** is a visual form of a hierarchical pattern (1.2.3) that relates organizations, or networks of organizations, in time.

1.15 **Depth:** is the dimension of mindspace that creates the layers of hierarchical patterns. It intersects our commonly known definition of the 3rd dimension, and is stored within logarithmic scales.

1.16 **Attraction:** is any force that brings circles closer together and reduces the length of lines in mindspace.

1.17 **Stack:** is any connected branch of organizations within a hierarchical pattern (1.2.3).

1.18 **Complex Arrangement:** is a circle containing interconnected circles that are usually described within a network diagram (1.2.2). Any organization is a complex arrangement with some unique c-ratio (1.9). All systems of matter are complex arrangements containing an abstract mass in mindspace.

1.19 **Scale:** refers to a level of depth. Organizations on higher scales have a greater size than organizations on lower scales, and they are found higher in the hierarchical diagram (1.2.3).

CHAPTER 2

2.1 **Energy:** is the sum of potential and kinetic energies within a closed system.

2.1.1 **Potential**: In thought theory, is the measure of size within a mindspace organization.

2.1.2 **Kinetic**: In thought theory, it is the measure of lines within a mindspace organization.

2.2 **Vis Viva:** is an early formulation of the principle of conservation of energy, which describes kinetic energy (2.1.2). It describes the transfer of potential energy into kinetic energy, and vice versa. It can be used to understand the elastic flow of energy across mindspace.

2.3 **Abstract Time Dilation:** An abstract form of Einstein's Time Dilation that arises from the interaction of patterns in mindspace. Abstract time dilation causes particular arrangements of matter to materialize faster within an environment than other potentially competing arrangements. It is affected by shifts in complexity or c-ratio (1.9).

2.4 **Pi:** is a ratio of a circle's circumference to its diameter. In this book, it is considered to be the basic measurement of motion in mindspace, aka. rotation. Motion in mindspace is used to concentrate resources within circles.

2.5 **Waves:** are a result of the periodic motion fueled by PI. Irregular waves are instances of rotating circles that are changing size and rotational speed. It is speculated that calculus could be redefined in terms of circles, rather than waves.

2.6 **Wave Dimensions: (cycle of a circle)**

2.6.1 Period (unit): proportional to any single cycle.
2.6.2 Amplitude (pitch): proportional to the scale of a cycle.
2.6.3 Wavelength (context): proportional to the size of the cycle. (size)
2.6.4 Frequency (repetition): proportional to the speed of the cycle. (complexity)

2.7 **e:** is approximately equal to 2.718. The number **e** can be applied in probability theory, relating to exponential growth. It is related to the growth in size of circles in mindspace.

2.8 **Phi:** When approximated its value is 1.618 (known as Phi). As a ratio, its value is 0.618 (known as phi). It is a special number because the inverse of 1.618 is .618. It also has several other remarkable features in mathematics, leading to peculiar squares, waves, and spirals. It has been identified throughout nature, architecture, art and biology. It is related to the complex arrangement of circles and it is linked to the principle of mindspace symmetry.

2.9 **Sierpinski triangle:** is a very basic fractal geometry related to Pascal's triangle. It describes an ideal form of hierarchical organization in mindspace.

CHAPTER 3

3.1 **Tetractys:** A triangular figure consisting of ten points arranged in four rows such that one, two, three, and four points are in the rows. It is the geometrical representation of the fourth triangular number. It is a base of Thought Theory, explaining the structure of mindspace as a connection between the decimal and duodecimal number systems.

3.2 **Pythagorean Theorem:** The most important equation in the Universe. It constructs the fundamental diagram.

3.3 **Hyperdimensional Space (HS):** is the overall architecture for the basic mindspace properties. It is primarily based on the union and hierarchical patterns. The HS arises from the sequence described by Pythagorean symbolism - starting from the monad and ending with the decad.

3.4 **Dyad:** is the basic structure of the HS. It joins separated monads into a shared environment.

3.5 **Triad:** is the extension of the dyad into the depth of mindspace. It represents the three principles of mindspace, and creates a mirror for measuring symmetry.

3.6 **Tetrad:** represents stability and equilibrium. It explains the numbers of dimensions in each group (physical, intelligence, and wave).

3.7 **Decad:** A group or set of ten; Pythagoreans viewed the decad as a special assembly and a symbol of earth and heaven. It can also be regarded as a junction between the decimal and duodecimal belief systems.

CHAPTER 4

4.1 **Sun Symbols:** Refer to 7.5

4.2 **Duality:** describes an encompassing relationship between two opposites. For example, a duality of the circle is defined by the circle's inside versus its outside.

4.3 **Abstract lifespans:** is measured by the periodic motion of symbols in mindspace.

4.4 **Universal:** is anything that two or more particular arrangements of matter have in common, namely qualitative or quantitative value. It is another word for symmetry, used by metaphysicists. For example, two arrangements might share the color green. Universals are hypothesized to have calculable effect on a mind and a society of minds.

4.5 **Irrational Constants:** are values that cannot be expressed as a fraction. The most important irrational constants are (in order) Pi, Phi, **e**, $\sqrt{2}$, $\sqrt{3}$, $\sqrt{5}$. These 6 values produce most of the framework for the hyper-dimensional architecture (HDS).

4.6 **Three Principles:** are the guiding philosophies of the three components of the circle that transcend into the three families of dimensions and the three irrational constants. Also known as the principles of order.

> 4.6.1 **Symmetry:** is a relationship that interconnects two regions of mindspace into a single fundamental mechanism.
> 4.6.2 **Attraction:** is a force that pulls interconnected regions [within relationships] closer together in mindspace .
> 4.6.3 **Negation:** is an effect of complexity that can break relationships and interfere with attraction, yet is also used to maintain relationships.

4.7 **Fundamental Diagram:** Represents an angle and its three components. It also represents the circle, and binds the circle, and angle, to three irrational constants (Pi, Phi, **e**) that provide the base mechanism for creating circles. It illustrates how the three principles of mindspace work together to forge interactions between circles within mindspace.

4.8 **Fundamental Mechanism:** Refer to 0.6. It is a chiral mechanism across the fundamental diagram. It binds concentration and interaction to an equilibrating mechanism that balances mind and matter in mindspace. It is also the source for the wavefunctions found in quantum physics.

CHAPTER 5

5.1 **Chaos:** is a state of randomness. It is without order and coherence. Chaos implies all lack of intelligible patterns.

5.2 **Order:** is opposed to chaos. By way of the fundamental mechanism (0.6), energy is ordered into sequenced arrangements, called complexity (1.10).

5.3 **Attraction:** Refer to 1.16. It arranges circles and lines into patterns of increasing complexity.

5.4 **Gravity:** is a curvature in the fabric of spacetime. Masses bend spacetime and nearby arrangements of matter, known as objects, fall into these wells. The wells, also called warps in TT, are created by the size of circles in mindspace.

5.5 **Electromagnetism:** is a quantized forced emitted by electrons. This force travels the magnetic and electric fields, and it is called electromagnetic radiation, or otherwise known as light, photons, or a type of spin 1 boson. Electrodynamics explains that electrons and photons are responsible for connecting arrangements of atoms together. E.M. is considered to be the physics governing the lines between circles.

5.6 **Concentration:** is the abstract equivalent to gravity. In a mindspace, a circle will absorb other surrounding circles into its set of resources, contributing to its size (1.10).

5.7 **Interaction:** is the abstract equivalent to electromagnetism. In a mindspace, circles will exchange size with other circles across their relationships. Relationships provide pathways to interact and exchange tiny circles, that materialize as quanta.

5.8 **Negation:** is the third law of mindspace that relates to the line, interference, and waves. It is used to explain destructive complexity, and concepts such as war, fear, pain, stress, and difficulty.

5.9 **Destructive Complexity:** is caused by the excess accumulation of complexity between organizations during attraction. Minds collide with nearby minds and as this happens, relationships breakdown and decay. Minds will break away from destructive complexity. It is analogous to destructive interference in electrodynamics.

5.10 **Warp:** is a region of mindspace with a higher density of complexity. Warps are regions with higher c-ratios and they offer accelerated (time dilated) pathways for travelling minds.

5.11 **Abstract Mass:** The component of mass that is a sum of abstract relationships between atoms, rather than being a sum of their individual weight. An abstract mass is a specific value of a warp.

5.12 **Periodicity:** is related to the orbit. From different perspectives, an orbit can be referred to as a sequence, curve and Riemann surface. In electrodynamics, it can be referred to as the simple harmonic oscillator. All arrangements have wave properties and behave routinely.

5.13 **Environment:** is any photograph. Anything that can be photographed, from any distance and within any scale, is an environment. Environments contain arrangements of matter, called symbols, and over time the arrangements will move positions. Different symbols appearing in the same photograph (environment) share some type of symmetry (1.8).

5.14 **Vesica Piscis:** is the center region of the dyad (3.4). It translates into 'Fish Bladder' in Latin, and it represents a Union. When two organizations are united, all of their existing shared environments emerge within the vesica piscis, revealing some other probability of a future intersecting event in some shared physical environment.

CHAPTER 6

6.1 **Energy:** is a quantity of time required to perform work. Energy is indirectly observed and is a function of the amount of work done. The total energy of a system is a value equating to all the work that it can perform in a period of time. Energy is stored within mindspace organizations.

6.2 **Mass:** is a measure of energy derived from a measure of matter. In physics, calculating the mass of matter is the principal means of calculating energy. Mass is a physical reflection of an abstract principle, called energy.

6.3 **Environmental shifting:** is the name given for the process of moving large quantities of protons and neutrons across a body. As something travels, it must push across matter and there is a universal resistance to this activity which can be overcome with a deeper configurations of mindspace patterns.

6.4 **Folding:** is the process of creating a Venn diagram (Union pattern) between time and space. This process is caused by the principle of attraction, and gives rise to the physical and abstract versions of its forces. This process gives birth to the fundamental diagram and its mechanism.

6.5 **Rotation:** the action of turning a system around an axis or center. Rotation and translation are the two basic motions across a 3-dimensional space. It's the top view of the motion of mindspace organizations.

6.6 **Translation:** A uniform and linear motion without rotation in which the origin of a coordinate system moves to another position without altering the direction of the axes. It's the side view of the motion of organizations in mindspace. Translation and rotation are the two basic motions across a 3-dimensional space and as mindspace organizations rotate, they are also translated into lines, releasing wave phenomenon in our physical world.

6.7 **Four Physical Forces:**

   6.7.1 **Strong:** is a nuclear force binding quarks and hadrons.
   6.7.2 **Weak:** is an atomic force causing decay.
   6.7.3 **Electromagnetism:** is a short and long range force between structures in atoms. (4.5)
   6.7.4 **Gravitation:** is a long range force between structures of atoms. (4.4)

6.8 **Four Abstract Forces:**

   6.8.1 **Concentration:** is related to gravitation. A force between organizations of varying scale.
   6.8.2 **Complex Interaction:** is related to EM. A force between organizations of similar scale.
   6.8.3 **Connect Time:** is related to strong force. A force binding circles and lines.

6.8.4 **Rotate Space:** is related to the weak force. A force causing circles to rotate along connected lines.

6.9 **Boson:** Any of a class of particles, such as the photon, pion, or alpha particle, that have zero or integral spin and obey statistical rules permitting any number of identical particles to occupy the same quantum state. They create lines that criss-cross mindspace.

6.10 **Fermion:** A subatomic particle, such as a nucleon, that has half-integral spin and follows the statistical description given by Fermi and Dirac. They create the distinct insides and outsides of circles in mindspace.

6.11 **Mental Economics:** Relating to the production, development, and management of material and mental wealth. It is hypothesized that all exonomic systems are basically effected by minds and their ability to manage patterns using the three fundamental principles of mindspace.

6.12 **Work:** Work is a force acting a length of space and is equivalent to energy exerted to pull or push against the forces of nature. In thought theory, growth, concentration (5.6), and supply are the necessary pre-requisite actions for eventually perfoming work. It is the purpose of rotation.

6.13 **Play:** Is the physical activity of performing work. This leads to a change, interaction (5.7), and demand. The idea of work is manifested during the work phase, while the activity is manifested during the play phase. It is the activity caused by rotation.

CHAPTER 7

7.1 **Euclidian Grids:** A space in which the axioms and definitions of Euclid apply. A metric space that is linear and finite in dimension. Mindspace is organized by grids.

7.2 **Logarithmic Scales:** A scale of measurement that displays the value of a physical quantity using intervals corresponding to orders of magnitude, rather than a standard linear scale. Mindspace grids are also organized by increasing scales of depth.

7.3 **OI Diagram:** Outside-Inside diagram describing the basic construction of the Mandelbrot Fractal. It is an advanced diagram based on the system pattern that will be elaborated on in a future book. It is another way of describing the fundamental mechanism (its top view).

7.4 Mind Matter Mechanism: is the probabilistic rules by which thought attracts matter. If minds work together, arrangements of matter will manifest in their environment, that is symmetrical to their patterns of thought.

7.5 Sun Symbols: The relationship patterns can be interpreted as diagrams of the Sun.

    7.5.1 **Network (Network Diagram):** Refer to 1.2.2
    7.5.2 **System (OI Diagram):** Refer to 1.2.1 and 7.3
    7.5.3 **Union (Venn Diagram):** Refer to 1.2.4 and 3.4
    7.5.4 **Gear (Mechanical Diagram):** Refer to 1.2.5
    7.5.5 **Hierarchical (Fundamental Diagram):** Refer to 1.2.3

7.6 Entropy: is the quantitative measure of disorder or randomness in a system. A thermodynamic quantity representing the unavailability of a system's thermal energy for conversion into mechanical work. The second law of thermodynamics says that entropy always increases with time. A logarithmic measure of the rate of transfer of information in a particular message or language.

7.7 Dimensions of Intelligence: Refer to 1.13

7.8 Physical Dimensions: Refer to 0.2

7.7 Wave Dimensions: Refer to 2.6

7.8 Vis Viva: Refer to 2.2

7.9 Inverse Square Law: a law in which the magnitude of a physical quantity or intensity varies with the square of the distance from its source. The magnitude of light, sound, and gravity obey this law, as do other quantities.

CITATIONS

Introduction

1. "Is there a more fundamental theory?", Official string theory website, http://www.superstringtheory.com/basics/basic7.html, (accessed December 20, 2012).

Chapter 1

1. Jowett, Benjamin, and Inc NetLibrary.Timaeus. Champaign, Ill: Project Gutenberg ;, 2006.

2. Blackburn, Simon. The Oxford dictionary of philosophy. (Oxford: Oxford University Press, 1994), 145-54.

3. Ghirardi, Giancarlo, "Collapse Theories", The Stanford Encyclopedia of Philosophy (Winter 2011 Edition), Edward N. Zalta (ed.), URL = <http://plato.stanford.edu/archives/win2011/entries/qm-collapse/>.

Chapter 2

1. Curd, Patricia, and Richard D. McKirahan.A Presocratics Reader Selected Fragments and Testimonia.. 2nd ed. (Indianapolis: Hackett Pub. Co., 2011), 73-100.

2. "Pi - Wikipedia, the free encyclopedia." Wikipedia, the free encyclopedia. N.p., n.d. Web. 15 Nov. 2012. <http://en.wikipedia.org/wiki/Pi>.

3. Soanes, Catherine, Sara Hawker, and Julia Elliott. Oxford dictionary of current English. 4th ed. (Oxford: Oxford University Press, 2006), 1044-1045.

4. "File:Sine curve drawing animation.gif." Wikipedia. N.p., n.d. Web. 17 Oct. 2012. <http://en.wikipedia.org/wiki/File:Sine_curve_drawing_animation.gif>.

5. "Complex Numbers and Simple Harmonic Oscillation." Galileo. http://galileo.phys.virginia.edu/classes/152.mf1i.spring02/ComplexNumbersSHO.htm (accessed December 20, 2012).

6. "Why do particle tracks curve?." Ask an Expert at able2know - Ask Questions, Share Answers. http://able2know.org/topic/168946-1 (accessed December 20, 2012).

7. "Periodic table - Wikipedia, the free encyclopedia." Wikipedia, the free encyclopedia. N.p., n.d. Web. 20 Nov. 2012. <http://en.wikipedia.org/wiki/Periodic_table

8. "Mayan Periodic Chart of the Elements." Mayan Periodic Chart of the Elements. N.p., n.d. Web. 10 Sept. 2012. <http://www.mayanperiodic.com/>.

9. "File:236084main MilkyWay-full-annotated.jpg." Wikipedia, the free encyclopedia. N.p., n.d. Web. 10 Nov. 2012. <http://commons.wikimedia.org/wiki/File:236084main_MilkyWay-full-annotated.jpg>.

10. "Orion." Online Etymology Dictionary. Douglas Harper, Historian. 20 Jan. 2013. <Dictionary.com http://dictionary.reference.com/browse/orion>.

11. "The brain a galaxy of neurons" The Fountain. Murat Sonmez, Issue 28 / October - December 1999 <http://www.fountainmagazine.com/ Issue/detail/The-Brain-A-Galaxy-Of-Neurons>.

12. Knott, Dr. Ron. " The Fibonacci Numbers and Golden section in Nature - 1 ." Mathematics University of Surrey - Guildford. N.p., n.d. Web. 20 Jan. 2013. <http://www.maths.surrey.ac.uk/hosted-sites/R.Knott/Fibonacci/fibnat.html#Rabbits>.

Chapter 3

1. Wehrli, Fritz. Die Schule des Aristoteles: Texte und Kommentar. 2. Aufl. ed. (Basel: Schwabe, 1978).

2. Guthrie, Kenneth Sylvan. The life of Pythagoras. (Alpine, N.J.: Platonist Press, 1919), 6.

3. Huffman, Carl, "Pythagoras", The Stanford Encyclopedia of Philosophy (Fall 2011 Edition), Edward N. Zalta (ed.), URL = <http://plato.stanford.edu/archives/fall2011/entries/pythagoras/>. http://www-history.mcs.st-and.ac.uk/Biographies/Pythagoras.html

4. Letter by letter, An alphabetic Miscellany
Laurent Pflugaupt 2007 Princeton Architectural Press

5. "Ancient Egyptian Writing, Hieroglyphs, Scribes - Crystalinks." Crystalinks Home Page. N.p., n.d. Web. 17 Jan. 2013. <http://www.crystalinks.com/egyptwriting.html

6. "Freemasonry ." Henri Masonic Lodge. N.p., n.d. Web. 2 Oct. 2012. <www.henrimasoniclodge.org/site1/Freemasonry.aspx

7. "Scottish Rite of Freemasonry, S.J., U.S.A.." Scottish Rite of Freemasonry, S.J., U.S.A.. http://scottishrite.org/ (accessed December 20, 2012).

8. Laertius, Diogenes . Diogenes Laertius: lives of eminent philosophers. (Cambridge, Mass.: Harvard University Press, 1980)

9. "How can E8 be the theory of everything? - Curiosity." Curiosity : Discovery Channel. N.p., n.d. Web. 17 Jan. 2013.
<http://curiosity.discovery.com/question/e8-theory-of-everything>. (248 dimensions)

10. Garrett Lisi, A., An Exceptionally Simple Theory of Everything. arXiv:0711.0770 (2007). Available at URL http://arxiv.org/abs/0711.0770.

Chapter 4

1. "Dimension." Dictionary.com Unabridged. Random House, Inc. 20 Jan. 2013. <Dictionary.com http://dictionary.reference.com/browse/dimension>.

2. MacLeod, M. C., & Rubenstein, E. M. (2005, December 9). Universals [Internet Encyclopedia of Philosophy].
Internet Encyclopedia of Philosophy. Retrieved December 2, 2012, from http://www.iep.utm.edu/universa/

Chapter 7

1. "Dan Dennett: The illusion of consciousness | Video on TED.com." TED: Ideas worth spreading. N.p., n.d. Web. 20 Jan. 2013.
<http://www.ted.com/talks/dan_dennett_on_our_consciousness.html>

·

THEORY OF THOUGHT